ALCHEMICAL MANUAL
for this MILLENNIUM

Vol. 1

ALCHEMICAL MANUAL
for this MILLENNIUM

Vol. 1

Aaity Olson, scribe

copyright©2020 Aaity Olson

New circumstances will reveal the path to the future. Truth is always the same, but combinations are different, as they depend on the consciousness. So much that is beautiful ends up being destroyed due to ignorance of the temple of the heart. But let us adamantly strive to be aware of heartfelt warmth, and let us start feeling ourselves to be bearers of this temple. That is how we can cross over the threshold into the New World. How inconsequential are the people who imagine that the New World is not for them. Bodies may differ, but the spirit cannot evade the New World.

Forward

I was friends with Aaity Olson for a short while before she left this world. I found a snippet of her work somewhere online and wrote to her requesting a full copy of her books. All of the digital copies of her work currently available are from the copy she sent me on CD in 2010.

These books are a wealth of technical and spiritual knowledge. I am publishing physical copies in order to preserve her work. The proceeds from the sale of these physical books will be used to fund research in symmetric field physics. The digital copies in her original format are and always will be freely available. If for some reason they disappear from the Internet I will reseed them.

I am forever grateful to Aaity for her work and her sacrifice.

Kelly Roman
Hyacinth Research
kelly@hyacinthresearch.com

Preface

The following is taken from emails Aaity sent me:

Hi Kelly,

Some years ago I started to think about everything in terms of energy flows, passive circuits and invisible tori. I see proportions and harmonics everywhere, especially in nature. I love music but I am not a musician. I am an artist with eyes to see inward and outward. At a conference a man asked me if I could see tori. I told him sometime I could see them and sometimes I could see the results of them, most often the results. And anyone can if they take the time to observe.

I believe it is important to recognize tori both because it would help in understanding of how things work on earth, and it would show people the reality and power of the S Field. It should also help females to grasp that their energy is important in the world and beyond. It will help people to realize that harmonics have spiritual structure. In reading the work of Hans Jenny I hope that you can see that it is the invisible tori that gather the materials on his plates and move them into visual patterns. You can trace the flow of motion tori around the currents in his dishes. Tori, motion in Ether, help to carry our messages in to outer space across millions of miles. Our thought and our prayers can be heard, even without a cell phone. Can you help to teach people about tori?

I grew up with Theosophy (like Keely and his buddies). Most of all I am interested in history and what ancient peoples believed. It came to my attention that ancient people were making artworks that showed they had a knowledge of the science told to me by the Star People. How did they know about the galaxy, the energy of space, the flight of people, the circuits that communicate? Well, from my experience the star People must have been around for a long time. If someone would listen, they would talk. Sit quietly and listen! Then put a call into outer space and include a question. But believe that they will tell you the truth (then check it out). They are apt to surprise you.

Les Brown made a passive circuit with coat hangers. By accumulating enough energy into a point potential he changed the predominance of the environmental field. (Something like the ark of the covenant did). But playing around with alternative reality a person may find what he does not want.

Aaity Olson

I am sending you some photos I took in Cartego, Costa Rica. The story goes that in Cartego, a long, long time ago, a young girl found a carved Madonna that was made of dark greenstone (Nephrite). She brought it home as a doll. But the statue kept disappearing and going back to her rock. Then the girl took the statue to the local priest and it did the same thing, over and over. Shut tight in a box, it kept returning to its rock. The priest got the message and built a church around the statue. But earthquakes destroyed the church a couple of times. (Costa Rica is full of volcanoes and earthquakes.) Meanwhile the people kept coming and asking for miracles and cures. Everyone was convinced that this carving was a point for miracles because word got around that it was true. People kept coming, so the Catholics kept building. The pictures will show the real statue, Her Church and the plaza that is filled with thousands of pilgrims every year. Just any weekday shows a steady stream of worshipers, some walking up the marble aisle on their knees. Here is a picture of the alter with the statue surrounded by a world of gold and precious symbols of church, angels, etc.

Here is a picture of the alter that, at first glance, makes one think She is the Sun Incarnate. Look again. Each wire is stacked with gold beads in this radial pattern. Every angle is there. The church architects were not going to take any chances. The antenna for their little doll were aimed right at the point potential of her power. She is a sitting radio from outer space. As many as 30,000 people every year pray before her with their families. They must be doing something right! I had chronic bronchial infections at the time. A year later I am cured of it. Of course, I MOVED TO FLORIDA.

Is this the only personage to be surrounded by gold antennas? Don't think so. What about the 80 foot wooden figure of Athena in the ancient Parthenon, clad in heavy gold leaf? I think the statue was filled with special alloys of metals. The barbarians cut her down. But you have to ask, WHO is out there?

Gold is more than pretty. It leads you into different realities.

Try Nathan Stubblefield's rods in the ground for a better crop.

If you meditate and walk inside an atom what do you see?

Enjoy Life

Aaity

January 24th 2010

Dear Kelly,

This letter is about ME. Maybe as an onlooker, you may see something of yourself, too.

In April I will be 78 years old. I have had, on the whole, a poor but rewarding life. My mind and body are getting unreliable but I keep pushing toward a genuine functionality. I am one of those old hippys who dropped out in the seventies to a cabin at 8000 feet in Colorado. The economic system prevented a financial stability but the experience was incomparably great for me. I lived all over the west and have gotten wiser with each year that goes by. I am a very good artist and what that means is that I see beauty everywhere and especially in nature. And, of course, the west provides beauty at every step on the trial. I have two grown sons, older than you, and each has benefited from my insistence to live as close to the wilderness as possible. They still have doubts about my sanity. I now live with my friend Lois (79) in Havana Florida, north of Tallahassee where we have seasons but it is relatively warm and not like digging out 5 feet snow in Colorado. Together we have made 4 trips into Mexico to study ancient history, especially the progress of Mother Goddess in the Americas since the earliest times. Lois used to teach geology in New Jersey and we met because of our common deep interest in rocks. I have been to Tucson gem show only 3 times. LOVE it. That's because we vibrate to stones as though they are music. Each stone has its own song and we try to find it. We have small rocks and specimens all over the house. I have read about and practiced Esoteric ideas since I was very young and Lois tolerates that. As you know, I have been channeling for years (about 50) and mediating whenever possible, using raja yoga techniques. I learned the computer in my late 60s when I had to put the book together and I had my first broken leg. I am still learning.

I tend to be a depressive character. Because I could see how bad things were I got worse. Over time I learned that there is always a balance of progress and regression and that I could choose my focus. Much of the time I had some sort of medication, but I never used drugs nor alcohol. That was because I was too poor and also because I could not channel if I was not healthy. I was different from other people! I missed the company of people but could not talk about their interests, household things or politics, sports, etc. I read all the esoteric books I could find and spent any extra money on them. My most revealing author was Alice Bailey, and her Tibetan channel. She was an offshoot of Blavatsky. From her works I could easily see that I was Sirian and on the 5th Ray.

I learned to balance my depression with ritual and with putting myself in beautiful surroundings and making deliberate choices about how to spend my life (at risk of the consequences). Sure, I went into my inner world which was filled with star people who could tell me the truth. Crazy? Who cares? But in doing so lost my family. Then Lois came along who claims to know nothing about that but who thinks I am OK. She can't remember her Other lives and star associations and yet I can see that we have previous times together. I help her and she helps me and its good, even now. I learned that people usually get depressive as babies often because the right hormones do not get to certain parts of the brain and the brain gets accustomed to functioning that way, sometimes because of parents who are depressed, sometimes because of circumstances of neglect or loss. Hormone imbalance needs to be corrected medically. Before recent medical drugs people used beer or whiskey, didn't help much in the long run. People died because they chose to.

I have a wonderful pet wire haired terrier that comforts me. We understand each other. And I make sure that we have a party at least once a week, get dressed up and treat ourselves some way.

There is still another possibility of a feeling of isolation and the pain of depression. It is for a person who lives as a stranger on this planet. Now the internet helps because those strangers from different star system are finding a remote community.

So, I love science, but the study of physics is now undergoing such a change that I can't read anything for sure. My star people say that my work is done. I can relax with my hobbies and walking on the beach. I still want young people to be alert to the myriad positive changes that are taking place about learning and about an energy shift. The internet helps with that too. One of my hobbies is digital photography which I adore because it is so painless and so miraculous. And it gives me an excuse to get out into the Florida parks and to the all beautiful places. I give thanks to be an American and have lived long enough to see these wonderful tools in my own hands. HD is thrilling to me. (I only wish that I had those tools when I was in Sedona and Seattle, etc.) I can recommend that you have a fine digital camera, good on the trail, and be able to set off on foot with 16 gigs. I have a little Nikon but Lois's Cheaper Sony is better. Again, the computers make my day. I will send you some photos to prove it. Spring is coming here but the little buds have been frozen on the branches. Colder than in the west this year. When spring breaks we will take the new Sony HD Camcorder out for a real trial.

I hope you have not been bored. I hope we can be friends for a long time. I think we have a lot to share. Look for some pictures attached from time to time.

And by the way, don't worry about levitation and new energy and world betterment, it's all out there being done as we speak. Our job is to understand it when it is finally revealed, then to tell everybody the good news. It is done! Our artistry is to vibrate at will to new energies that are steadily coming into the planet. Yes, learning to vibrate with stones helps in the training and lets us talk to Mother Earth. Take a drive to Tumacacori and with your new camera and love the old and rejoice in the new.

LOVE and Light

Aaity

February 16 2010

Table of Contents

INTRODUCTION to ALCHEMICAL MANUAL for this MILLENNIUM, Vol. 1 7

CHAPTER 1
    SYMMETRIC FIELD PHYSICS 12

CHAPTER 2
    SYMMETRY IN MAGNETISM 33

CHAPTER 3
    THE CREATIVE BREATH OF GRAVITY 49

CHAPTER 4
    A NEW WAY TO SEE 85

CHAPTER 5
    VORTEX 103

CHAPTER 6
    WINDS OF QUETZALCÓATL 123

CHAPTER 7
    OPENING THE PETALS OF COMPASSION 147

INTRODUCTION to ALCHEMICAL MANUAL for this MILLENNIUM, Vol. 1

Here is an easy alternative science textbook describing a NEW concept of COSMOLOGY and an all-new <u>Symmetric Field Physics</u>; It introduces a revolutionary concept of reality that is both new to the modern world, and more ancient than recorded history.

The text is simply presented (no math) for all those people interested in the truth of the world around them. It is designed for students (young and old), fascinated with the physical world, who have found problems with the current theories taught in the academies of today. The new information presented by the Star People through scribe, A. Olson, is destined to overturn both classical physics and particle physics . Described as both revolutionary and profoundly ancient, the ideas within the text will illuminate a path for humanity to move forward in knowledge and promise to integrate all aspects of scientific studies.

Here is an astounding, life changing book explaining the LAWS OF COSMIC FORCES in SPACE, in detail, in both their physical and philosophical aspects. Vol. 1, printed in its entirety, is a gift to the world from the invisible Star People who wish to communicate with earth at this time.

Here is a brief synopsis of the contents of ALCHEMICAL MANUAL for this MILLENNIUM, Volume 1.

Symmetric Field Physics is the study of natural law and cosmic forces. The forces of space are not just up in the sky, they are all round you and through you. As a fish lives in the oceans, you live in a sea of powerful space. Nations and individuals are on the threshold to learning to use the energies of space.

Lessons presented about forces are chosen to review the classic experiments of the physics classroom with a new perception. They radically revise the old textbook analysis to show that altogether new ways of recognizing energy laws solve problems in a better way. These laws cover basic engineering, cosmic energy balance, theoretical physics, and the practical study of magnetism, electricity, gravity, weight, essentials of mechanics, weather, and the way natural growth takes place. The SPACE / TIME Field energy systems are discussed to enable a person to envision interfunctioning octaves of energy and the way octaves connect through vortical points of intersection within corpuscular fields that fill all space and time.

Emphasis is placed on the ever-present space fields, unseen by human eyes, but active as tori around every motion or action in the time fields. These space tori are both mechanical and electrical, covering a full spectrum equal and opposite to the linear activity of the corpuscular time field (electromagnetic field).

It is important to say that the text is designed to be read by a person of highschool level training with only basic science as a backround. It is wise to have a high school (or college) physics textbook handy as a reference. It was necessary to go back to the roots of physical studies in order to correct basic assumptions. Nevertheless, the studies are astute and thought provoking, often more challenging to an adult than to a youngster.

These ideas will be new to the student, therefore, it is wise to study the manual sequentially. New concepts and new words will have to be assimilated in order to be understood. Patience and review will reveal even more with each reading.

Here are some questions that the ALCHEMICAL MANUAL will answer. **(these are not the same answers that you were taught in school)**

1. What is magnetism?
2. What is electricity?
3. What is gravity?
4. Why do objects of different weight fall at the same rate?
5. What is momentum and inertia?
6. Why is the work of Isaac Newton incomplete?
7. Is there any empty space?
8. What is the nature of time?
9. How does space relate to time?
10. What are universal forces, and why are they invisible?
11. How is the galaxy structured?
12. What drives the weather systems?

And hundreds of other important questions about forces that influence your life.

ALCHEMICAL MANUAL FOR THIS MILLENNIUM

AAITY OLSON, SCRIBE

A MESSAGE FROM THE STAR PEOPLE

We are a group of star people authorized at this time to bring to earth vital information about the physical conditions in which you live. Even though you have made much scientific progress you cannot move forward unless you are willing to courageously examine new information that may not match your learning, and to make fundamental changes in your outlook regarding the nature of reality. Changes start at the beginning. You have been told to view the world in a particular manner when you were only a child, a manner dictated by your culture. Your teachings held erroneous information. Therefore, we dedicate this manuscript to the young men and women who may find ways to recover a more accurate concept of life on earth.

Times have changed. You see around you more evidence of how the Universe works. Education and experience for every person on earth has vastly increased. Still, scientific information has not been available to many people because they do not possess the mathematical skills to read and make sense of documents written in algebraic jargon. They, therefore, dismiss the study of physics as "too hard" to understand. That should never happen. Most anyone can learn the Universal Laws about energy. Complex as they may seem, the laws are consistent, and can be basically grasped by a ten year old child. To make that understanding possible very little math has been included in this book.

We are assuming here that the reader of this book has an open mind and has some basic interest and experience in general science. He or she may be an average person with a curiosity about how and why things work the way they do. An advanced student may be disadvantaged by previous assumptions. We beg their patience. The first chapter is, perhaps, the hardest to understand in the whole book because it deals with abstract principles. We suggest that you skip over what seems vague to you and go on reading, then come back to the laws over and over again, as you see how they might be applied. Not all the answers are given at first. This is a survey study and introduces one idea at a time.

Some of the ideas in this book are revolutionary to your thinking. Some are common knowledge. If you can grasp a few of the major premises in the work it will change and expand both you and your world. It is the ideas that are important, not the source.

At the start, we want you to know that the star people are real living beings that are a lot like yourself. We wish to communicate with you. Therefore, we choose to stay neutral in terms of names and places. Our earth contact, Aaity Olson, is not privileged to special information concerning us. Our secretary does not ask personal questions, nor should you. Our scribe is unschooled in science but she has outstanding artistic and spatial ability, and can see the visions we project of three and four dimensional diagrams. In this text you will see that there is certainly a language problem between old concepts and new. Bear with us, and you may have an opportunity to add your own perspectives to interpretations, especially regarding the definition of words. Name ideas in a way that all can understand.

We will ask you to work with energies that are invisible to you. Even as they are invisible you can see their shadow effects. Be patient with us and with yourself and all will be clear to you over time. Read as much as you can, then rest. Let the concepts be taken into your center.

It could be helpful to you to secure a basic old fashioned textbook on physics to use as a reference for statistics and to notice changes in concepts presented in this paper. Many of the old formulas are useful, while they cannot explain the physical events that take place in the fields at large. All the previous research is valuable, yet you will see that it will need to be understood in a new way using new techniques to enhance the usefulness of the work.

To probe and fathom invisible energies the scientific dictums of proof used in the past will have to be incorporated into a more flexible mindset with greater variables in the data bases. New computers will help you to deal with a great multiplicity of facts. Keep in mind that the details should never cover up the simple laws of truth. All people of all ages should have access to understanding the nature of the reality in which they live. Obscurity and elitism must be set aside. We will continually work with truth seekers who can realize that truth is for everyone. We come to you as brothers and sisters in the hope that you, who are reading this, will also be brothers and sisters to all.

> **The information is free.** The cost of information distribution is substantial. We will welcome contributions from our readers. It is requested that you join with others who welcome this information and coordinate appropriate research in order to bring to the whole world a new and old cosmology. Visit the website, download, buy and distribute the published works as often as you can. More information will be forthcoming in various media.

AN APOLOGY FROM THE SCRIBE

The messages in these volumes have been received by conscious channeling since 1961. However, they have been compiled as a written study since 1994. My only references have come from the public libraries. The work has been difficult and undoubtedly contain mistakes of a technical nature. I now feel a pressure to get this work to the public, with or without mistakes, with or without the elegance of a superb manuscript, as time seems to be of essence. I beg your forgiveness for my imperfections. The star people tell me that it is up to the reader to use the cosmology as presented in a way that will enhance life. The applications of Symmetric Field Theory are buried within the survey texts for those with eyes to see. As many cosmic mysteries are uncovered, many more mysteries come to the surface. The nature of ALL THAT IS can not be fathomed except as one small fallen leaf at a time.

CHAPTER 1

SYMMETRIC FIELD PHYSICS

IN THE BEGINNING

Our Universe is made up of field energies arranged in harmonic octaves. The greatest of these octave energies is the realm of GOD, a single point and continuum, also known as ONE spirit, or ALL THAT IS, or MIND, or any name peoples choose to honor the totality of life.

From this singular MIND of power comes a system so perfect, so mysterious, that none of us can fully understand it. This grand LAW OF CREATION begins with a simple division of energy, a point and a sphere: respectively, a contracting force and an expansive force.

DIVISIONS IN ABSOLUTE ENERGY

The division of ONE is now in creation. We live in a ONE self-creating, self-dividing Universe, among a multitude of Universes. No thing, no action, no life, no idea is separate from the ONE UNIVERSAL TIME-SPACE point and continuum as it emanates in Divine Ba lance. All is a function of ONE in symmetric expression. From this system arises the study of Alchemical Law.

One has created a spatial sphere that continues to radiate, and a center point that continues to contract. This is the first division of the ONE. The second division shows the Grand Sphere divided into two halves, each with a compressed sphere and a center point. The continuum divides itself continually in halves with center points and radii paths that are all interconnected. Each division is an octave, a group of resonant speeds.

The first division shows the points and the sphere to be equal and opposite in direction and in force. The idea of equal and opposite can be conveyed by this symbol; >< . Every subsequent division will be equal and opposite. Each half is joined together with a fulcrum of power. (see **Illustration 1-1**).

ILLUSTRATION 1-1

THE PROGRESSIVE DEVELOPMENT OF THE SYMMETRIC FIELDS

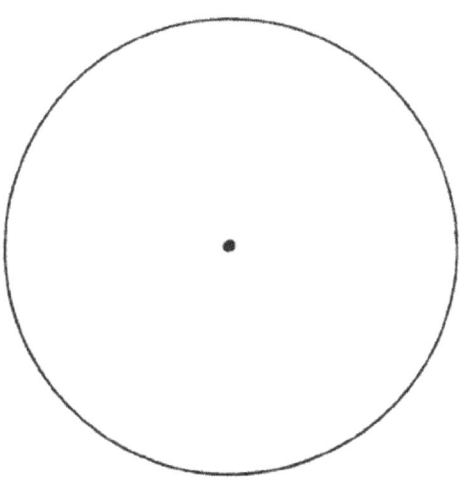

FIRST DIVISION OF **ONE**

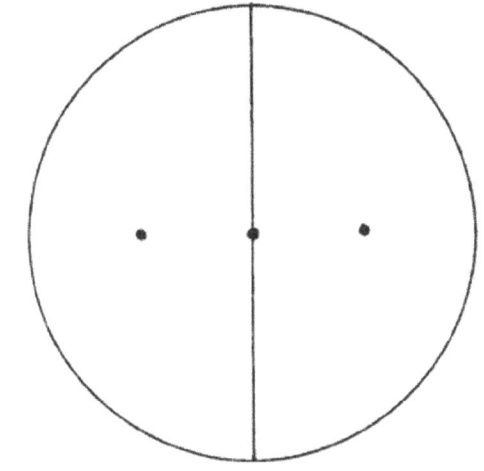

SECOND DIVISION OF **ONE**

EACH DIVISION CREATES EQUAL AND OPPOSITE HALVES
HALVES IN PARTNERSHIP

THIRD DIVISION OF **ONE**

FOURTH DIVISION OF **ONE**

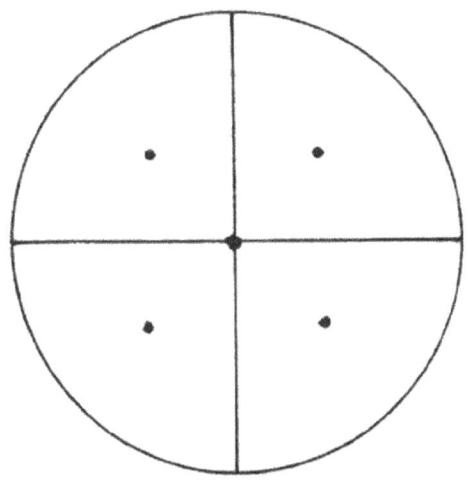

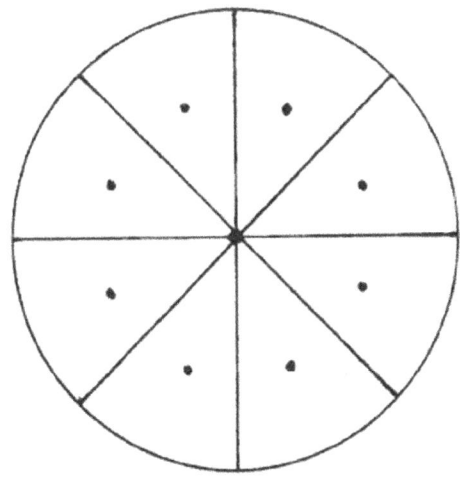

EVERY PART OF **ONE** CONTINUES TO DIVIDE IN HALVES
THAT BEST DIVIDE THE SPHERE INTO EQUAL VOLUMES

SPACE FIELD = TIME FIELD

There is an expanding continuum which we shall call SPACE FIELD (designated as S FIELD). It is the kinetic expression of MIND. It forms the sphere.

There is a contracting continuum which we shall call TIME FIELD (designated as T FIELD). It is the potential expression of MIND. It forms a point.

The SPACE FIELD acts equal and opposite to the TIME FIELD. The TIME FIELD acts equal and opposite to the SPACE FIELD. There is never an action in one without an equal and opposite action in the other. The fields stand together as SPACE-TIME FIELDS (designated as ST Fields). Although the fields stand together, they serve as an absolute barrier to one another. Each division in half creates a pair in the ST Fields. (see **Illustration 1-2**).

One S Field sphere with a T Field center divides itself into two spheres with centers. Each new sphere is a half of a pair. One half is S Field predominate, the other half is T Field predominant, even though each has an S Field sphere and a T Field center. Although this is confusing, it is an important part of the symmetric field system. Divisions of the grand spheres continue until unimaginably small paired corpuscles form an unbroken continuum.

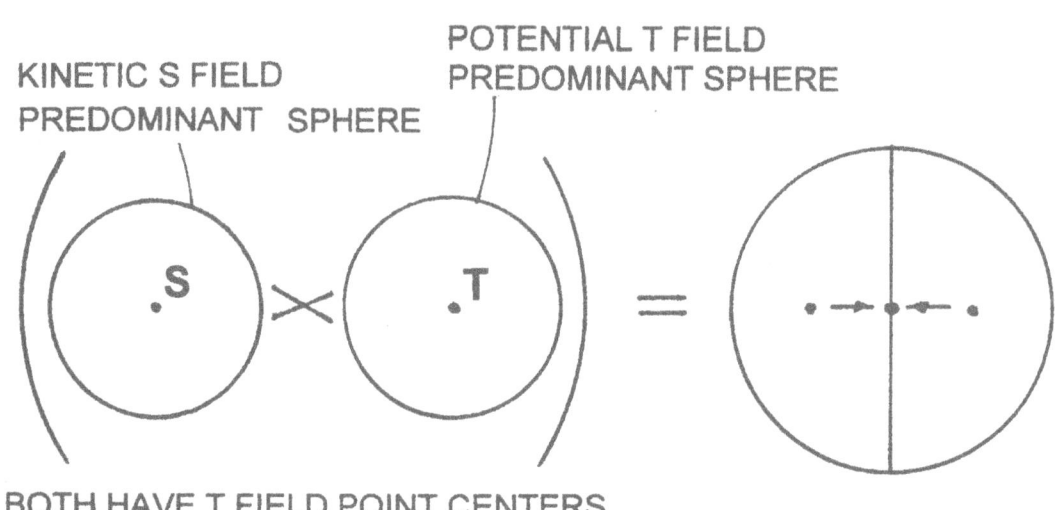

PREDOMINANCES IN FIELD SPHERES

ILLUSTRATION 1-2

The SPACE FIELD expands itself in all directions outward with equal pressures to form a sphere. It radiates from a T Field center point outward, to a diameter which exactly balances the potential energy in the point. At the limit of the circumference of the sphere, it begins to circle. At its maximum speed it reaches a "point of change = 0 ", a place where the S Field turns itself into T field. As T Field, it becomes a line and travels back to center.

The TIME FIELD moves from a spacious position to a plane, to a line, to a point. It attempts to form a dimensionless point, yet its energy is held at a PLACE for a DURATION. The T Field contracts in on itself. As the T Field lines travel to center, they slow down, begin to coil, and reach a center limit, relative to its "point of change = 0 ", where it changes to S Field and begins its outward movement once more. This is true for the gravitational forces around the earth and all other cosmic bodies.

The ONE expresses periodically in an in-breath and an out-breath, alternating the symmetric aspects of the ONE force. (see **Illustration 1-4**).

> **The S Field accelerates outward in all directions to the limits of its sphere. (expansion)**
>
> **The T Field decelerates inward to a line, then a point, to the limits of its speed. (closure)**

ILLUSTRATION 1-3
AXES AND RADII WITHIN SPHERES

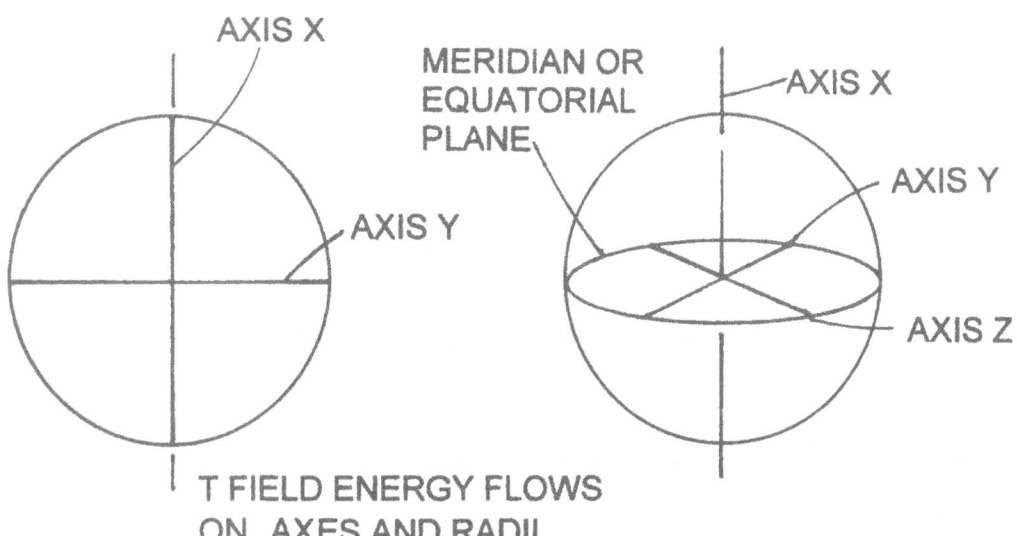

The S Field always repels itself in a radioactive thrust. We shall call this EXPANSION or RADIATION. Example: gas filling out a balloon. The T Field always attracts itself, contracting to the smallest possible place. We shall call this T Field contraction **CLOSURE**.

It is in these attributes we find all aspects of vector forces and material accretion.

The S Field and the T Field will not fuse, but will always stand side by side in the same octave. Any action in one field will mirror itself in the adjacent field. The two field energies balance on a fulcrum between the pair of corpuscles. There is no place where **pairs** of ST Fields do not exist. **There is no emptiness and no chaos.**

Although two opposite corpuscles stand side by side, they do not blend. They are absolute barriers to each other. By forming bonds in any number of configurations they develop boundaries, forming units that you recognize as substance. Any group of ST Field corpuscles that form a bond will develop an overtone sphere that envelopes the grouping. (See **Illustrations 1-2, and 1-5**).

It must be understood that attraction and CLOSURE describe a principle law whereas "repelling" does not. The S Field expands itself by internal pressure outward to enlarge the diameter of a sphere. When water is turned into pillows of tiny bubbles, for instance, it occupies greater dimension but it is not repellent by definition. T Field does not push S Field away but demands to stand side by side to S Field in balance. A T Field point or line is always surrounded by a swirling S

ILLUSTRATION 1- 4

INBREATH (T FIELD) AND OUTBREATH (S FIELD) IN ONE

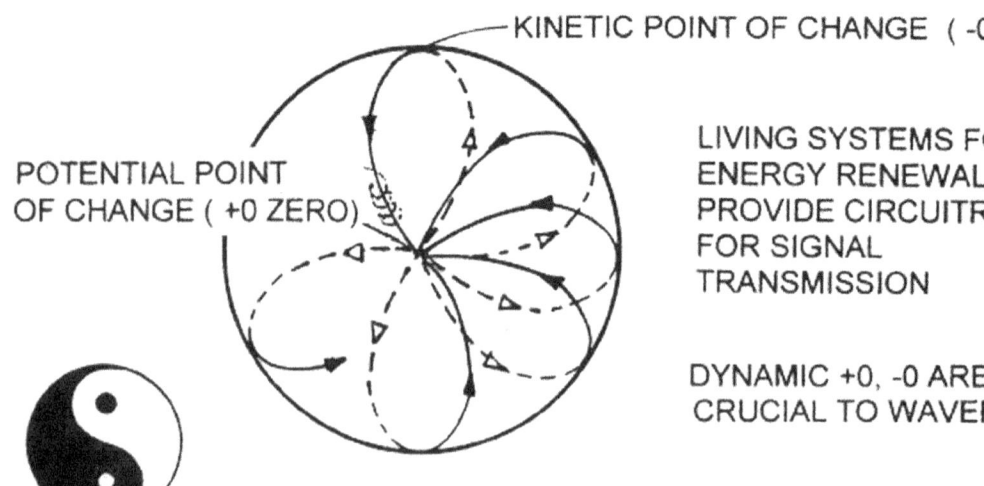

KINETIC POINT OF CHANGE (-0 ZERO)

POTENTIAL POINT OF CHANGE (+0 ZERO)

LIVING SYSTEMS FOR ENERGY RENEWAL PROVIDE CIRCUITRY FOR SIGNAL TRANSMISSION

DYNAMIC +0, -0 ARE CRUCIAL TO WAVEFORMS

Field sheath. The two fields form a barrier of definition between them. If one field comes forcefully against the other, it will be stopped and diverted, not repelled or annihilated.

Example: Electrons, which are T Field predominant units, traveling in wire are confined in that wire pathway by sheaths of S Field predominant rings in materials known as insulators. S Field defines T Field and vice-versa.

> **The Time Field is substance**
>
> **The Space Field is substance.**
>
> **The fields bond together to form many varieties of substances.**

Time as a substance is an idea which may be hard to grasp at first. T Field is the bonding force in the design of structures. The T Field in structure also "spends" time in cycles of duration . You recognize how materials "spend" time as you measure "half-life", the expulsion of mass from a material. Time you "spend" every day is an energy currency. Your DNA has units of T Field designed into its immaculate structure that express in cycles of living flesh and bone. Coordinated cycles of

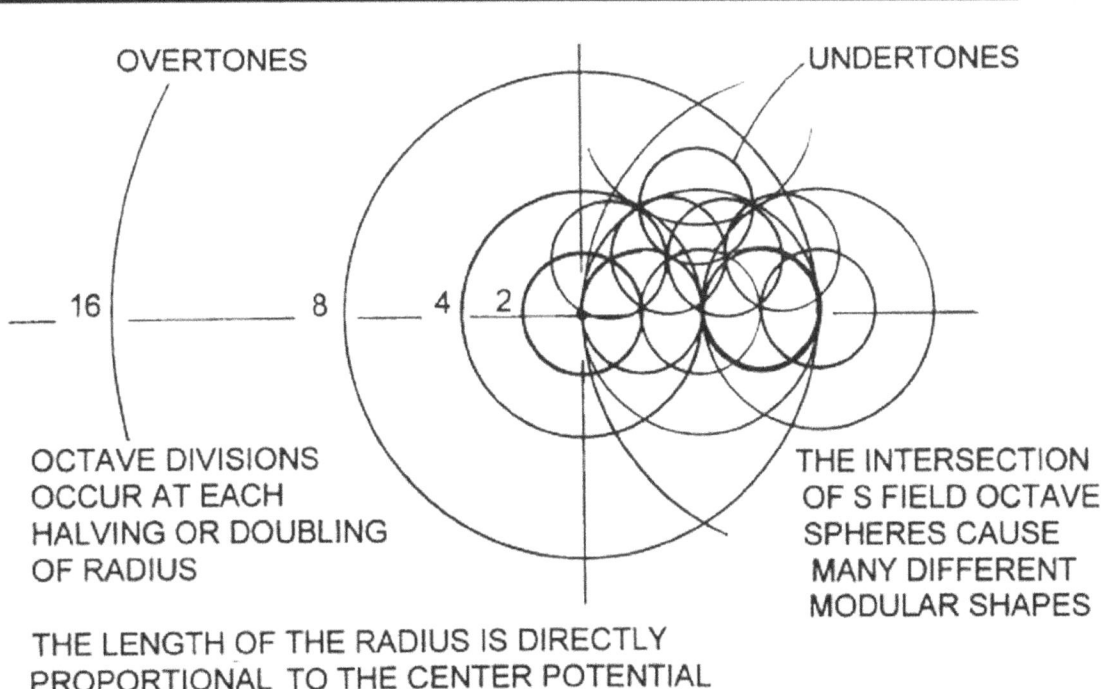

ILLUSTRATION 1- 5

OVERTONES

UNDERTONES

OCTAVE DIVISIONS OCCUR AT EACH HALVING OR DOUBLING OF RADIUS

THE INTERSECTION OF S FIELD OCTAVE SPHERES CAUSE MANY DIFFERENT MODULAR SHAPES

THE LENGTH OF THE RADIUS IS DIRECTLY PROPORTIONAL TO THE CENTER POTENTIAL earth sustain your being. You work under the sun, sleep under the stars, eat as the seasonal plants allow, etc. Every action requires both time and space energy. Time uncoils from its dwelling place of life to stimulate the regular beat of your heart.

> **The Space-Time Fields come together to formulate action in cycles.**
>
> **Energy and materials are the same entity.**

The substance of ST Field corpuscles are so fine as to express as a continuous ocean of energy, the fields pass through you like you were a sieve. The ST Field does not act as a barrier to you because **you are of it**. You swim in it like a fish in a sea. When the ST Field becomes very compressed, very dense, and tightly bonded it appears as a solid. Then it is defined by barriers.

Time is determined by the master clocks of the galaxy and the solar system. Yet each living thing experiences time in his/her own way. Time is both outside you and inside you. You coordinate many cycles manipulating a few octaves of energy. Standard time is a convenience worth having, but it is best to realize that the expression of time is quite individual and **absolutely relative to speed**.

As T Field slows down its linear bonds tighten. It becomes cold and hard and then its time extends into duration.

The universes are expanding according to their cycles. Stars move apart as **new energy** penetrates the system. As T Field centers gain more speed from travel and gain more potential from incoming energy, the S Field enlarges its spheres and speeds up to cover the distance. Because of this cycle, time expresses directionally for you.

Your personal sphere expands and memory is included in the duration of your sphere. Your NOW moments work out on the circumference of your sphere. The larger your sphere the more ideas can be related to bear upon your NOW. A person who sees the future makes a momentary, energetic enlargement of his/her sphere, often because of great personal concerns and empathic feelings. Your personal realization of time depends upon your own enlarged and elastic sphere.

> **Any adaptive process requires spherical flexibility and practices to increase personal potential.**

S-T FIELD IS USUALLY DEFINED BY ITS SPEED

In a galaxy like yours you have a fixed (average) speed base. Your solar system overtone provides a speed base. Rotation, travel, personal action, vibratory rates, metabolic rates; all things traveling together gives a resultant average speed rate which for you is much higher than you might want to realize. Even if the room you are sitting in is traveling with you at thousands of miles per hour each distinct material in your room has its own atomic vibratory rate. Very small differences in speed show great varieties of material manifestation. You should not presume that other planets in other galactic locations have materials just like your own.

However, the basic speed of a solar system does not completely limit the energy octaves available in that place. Octaves overlap as overtones and undertones. The suns and planets that you see are overlaid with suns and planets that you cannot see. Two or more things **DO** exist in the same place at the same time.

If you can look at the range of speeds within the known electromagnetic system and imagine an octave jump of the entire range, then you can conceive of the adjacent and overlapping energy system that should have your attention.

> **There is the difficult concept of speeds relative to zero where the quality of zero varies. A compendium octave change means a system change. That system shift is accessed through a harmonic geodesic concordance.**

We Star People are not much different from you Earth People. We use your octave of energy, but also enjoy the usage of the greater system. Some of our people were founders of your ancient civilizations. We were able to adapt to your solar system in the physical sense, while there was mental and emotional suffering at that time. The octave systems higher than our own are the planes where our God dwells, to Whom we are obedient. Power is given to those entities whose WILL is the same as the Will of God. That is a statement of the law of physics, not an ideology.

Measurements of cold and hot are measurements of field speed. When your fingers touch ice you say the ice is cold. In fact the ice is taking your heat for itself and storing it. The thermometer drops when put on ice. That means the mercury in the thermometer contracts when put on the ice. That means the mercury in the thermometer is contracting into **'closure'**. The scale reads a lower number. Contraction shows a slower speed and higher potential. Hot indicates expansion, a higher speed and greater kinetic energy. Hot and cold can measure octave energy states. Water can express in three octaves: Liquid, gas, and frozen solid. These octaves are within one system of octave divisions.

AXES ARE INTEGRAL TO OPPOSITE FIELDS.

Axes develop within an energy sphere regardless of the size and system. These axes locate relative to the overlaying fields as a whole, the travel within those fields, and the alignment of the fulcrum of the spherical groupings. They constitute an agreement between equal and opposite halves of a whole.

Axes are not imaginary, but actually present as highways of energy travel. The primary axis, (X), is a line defining a full diameter through the center point of the sphere. The secondary axis, (Y), is perpendicular to the primary axis and is a diameter through the center of the sphere. The tertiary axis, (Z), is perpendicular to both the primary axis and the secondary axis and defines a diameter through the center of the sphere. Together the Y and Z axes define an equatorial plane in the sphere. Radii act as a multitude of axes. Primary axis X combines with radii to form a multitude of meridian planes. These axes and planes are avenues for complex circuitry, (see **Illustration 1-3**).

THE NATURE OF THE CORPUSCULAR FIELD

Any ST Field divides itself into halves, (undertone octaves) along axial lines. Any ST Field doubles itself into twins (overtone octave) along axial lines.

Each spherical division has a place and an "address in space". Each corpuscle is individual but has precise heritage. Corpuscles do not move about unless they are disturbed by current winds and potential imbalances. There is not a fixed amount of energy. New energy is "born" every day. No energy is lost, but it has the ability to change its octave and associative relationships. Overtones can CREATE undertones. Overtones can absorb undertones. Undertones, by themselves, cannot effect overtones. In groups and harmonies, undertones can alter overtones. One octave of corpuscles tend to interact within the octave. It is their "comfort zone" (see **Illustration 5**).

It is convenient to think of musical tones in order to understand octaves. It is easy to play with harmonies between middle C and high C. The same relative harmonies can be found between middle G and high G. An octave greater than the octave of C would include low C to high C, (the spherical radius is doubled).

CORPUSCLES DIVIDE HARMONICALLY

As a radius line intersects an interior surface tension of its sphere, it creates a new center point on the circumference. This new center, in turn, spreads a sphere with radius equal to the first radius. Where the new sphere intersects the original sphere, it describes a cord and a plane. That cord harmonically repeats itself again and again around the sphere (all dimensions). The resulting figure describes a sphere with twelve sides. A cross section reveals a hexagram (see **Illustration 1-6**). The hexagram crystalline patterns are frequently seen in earth's materials.

As radii follow the lines of the axes they intersect the circumference of their sphere in the same way, forming cubes and figures with eight sides. Cords are formed at the axial intersections with the circumference. Many of earth's crystals are formed in this way. Chapter 11 deals with the way crystals are formed with their multiple harmonies.

Corpuscles fill all space. Although any corpuscle of the ST Fields are designed to be spherical, they are crushed together like bubbles in a basin. They take many shapes, most commonly, the cube and parallelepiped.

The corpuscles arrange themselves in alternating predominance's. They are locked in compression with potential strength according to their location. (see **Illustration 1-7**)

Both internal and external pressures can be understood in numerical harmonics. The relationships of chords yield stability – instability, time cycles, force augmentation and reduction, and many other geodesic co-ordinations including frequencies. Crystalline forms are worthy of careful study.

NETWORKS

> **The LAW states that any T Field line or plane, Intersecting another T Field line or plane will form a potential point of radiation at the place of intersection.**
>
> **The combined potential contribution to that center point develops an appropriate sphere with axes.**

The primary axis of a sphere is where current prefers to travel on the surface tension of a sphere.

There are T Field lines between every T Field point of similar magnitude. These pathways can be called circuits because they join with each other and also form a "ground" to a major center of ONE. (see **Illustrations 1-6, 1-7**)

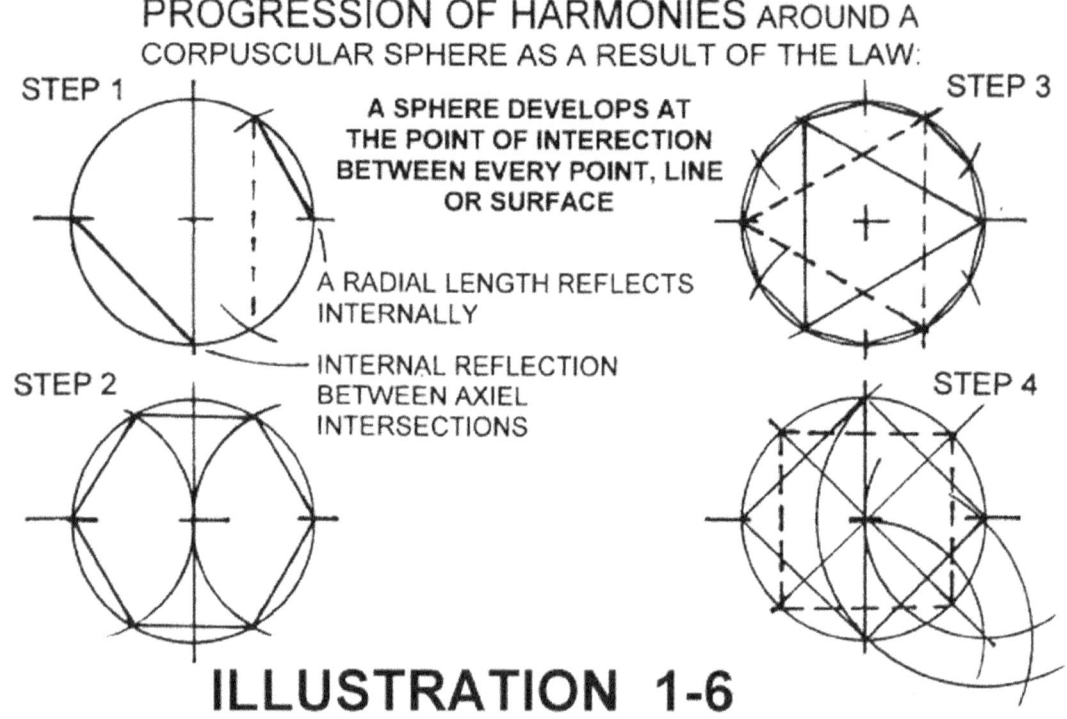

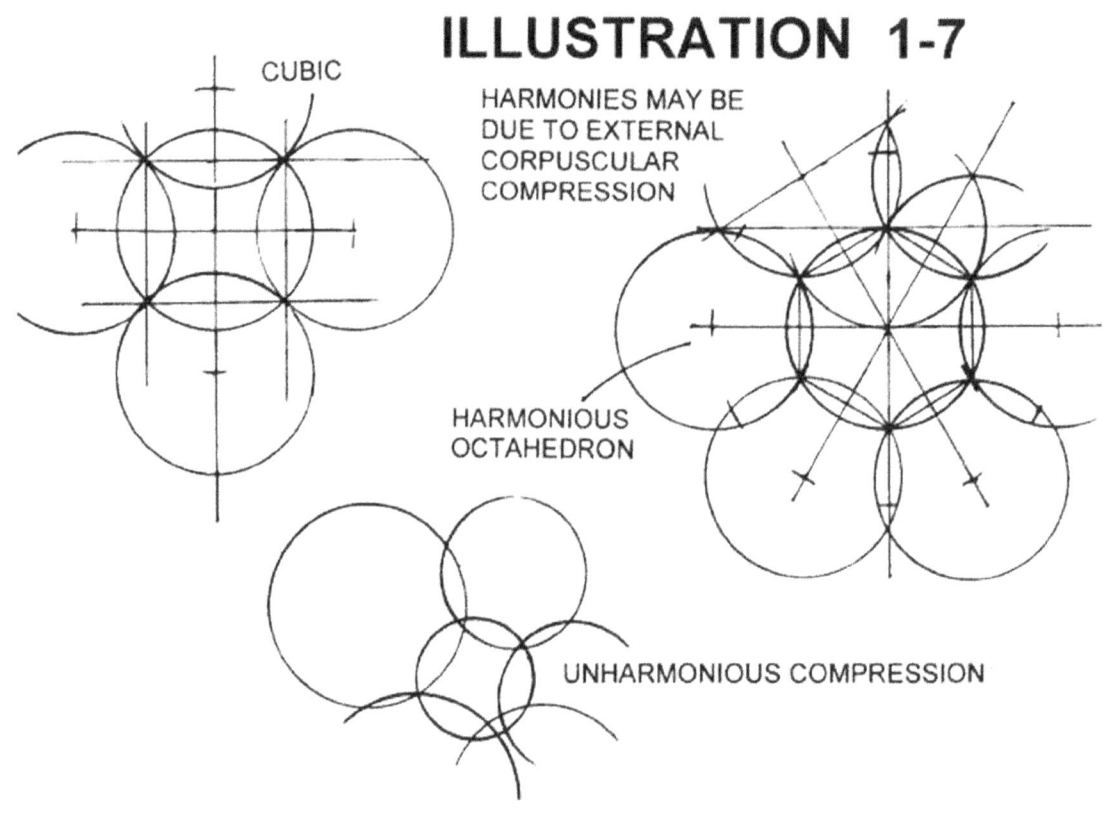

THE CREATIVE UNION

When two spheres of the same octave intersect, they form a common plane. Wherever the position of the two spheres may interlock, the common plane is always perpendicular to the line between the two centers of the spheres (radii). It is upon this plane that a union of energies come together to bond matter. It is upon this plane where both the lower and the higher fields merge with MIND; that both planning and materialization formulate along axes in motion. **Illustration 1-8** demonstrates the development of planes on an intersecting sphere. This subject will be detailed in our chapter on crystal formation.

Harmonies progress within a sphere as radii and other internal signals step in cords around the interior surface tension of the sphere.

> **A sphere develops at the point of intersection between every point, line, or surface.**

A newly developed sphere thus intersects and connects to develop new points of intersection in an unending fan of overtones and undertones.

CORPUSCLES YEILD TO COMPRESSION AND COMPASSION

Spherical corpuscles, alternating in predominance, fill all dimensions. There is no empty space.

Each corpuscle is compressed by neighboring external corpuscles in any harmonic or enharmonic configuration. **Illustration 1-7** shows common and uncommon planes resulting from field compression. The planar division of a sphere necessarily imparts an harmonic condition. Compression harmonies are translated into characteristic materialization. Any atom demonstrates its own peculiar harmonic, configurative characteristic. An atomic harmony cannot be identified by its "parts and pieces" but by its internal and imposed harmonies and can therefore act as a seedling for further building of atoms of its own kind.

If you bang about on an atom, putting it in an unnatural force field you will probably alter its harmonics. In this way you cannot know its nature. Take for example, a pet dog. If you are a mean master and beat on the dog, the dog will never reveal to you its personality. If you want intimacy with the dog, you will persuade it and listen carefully to its happy moments. An atom is a living, delicate instrument of life, just as is your pet dog. To understand it, be a loving master. Here is the first law of alchemy

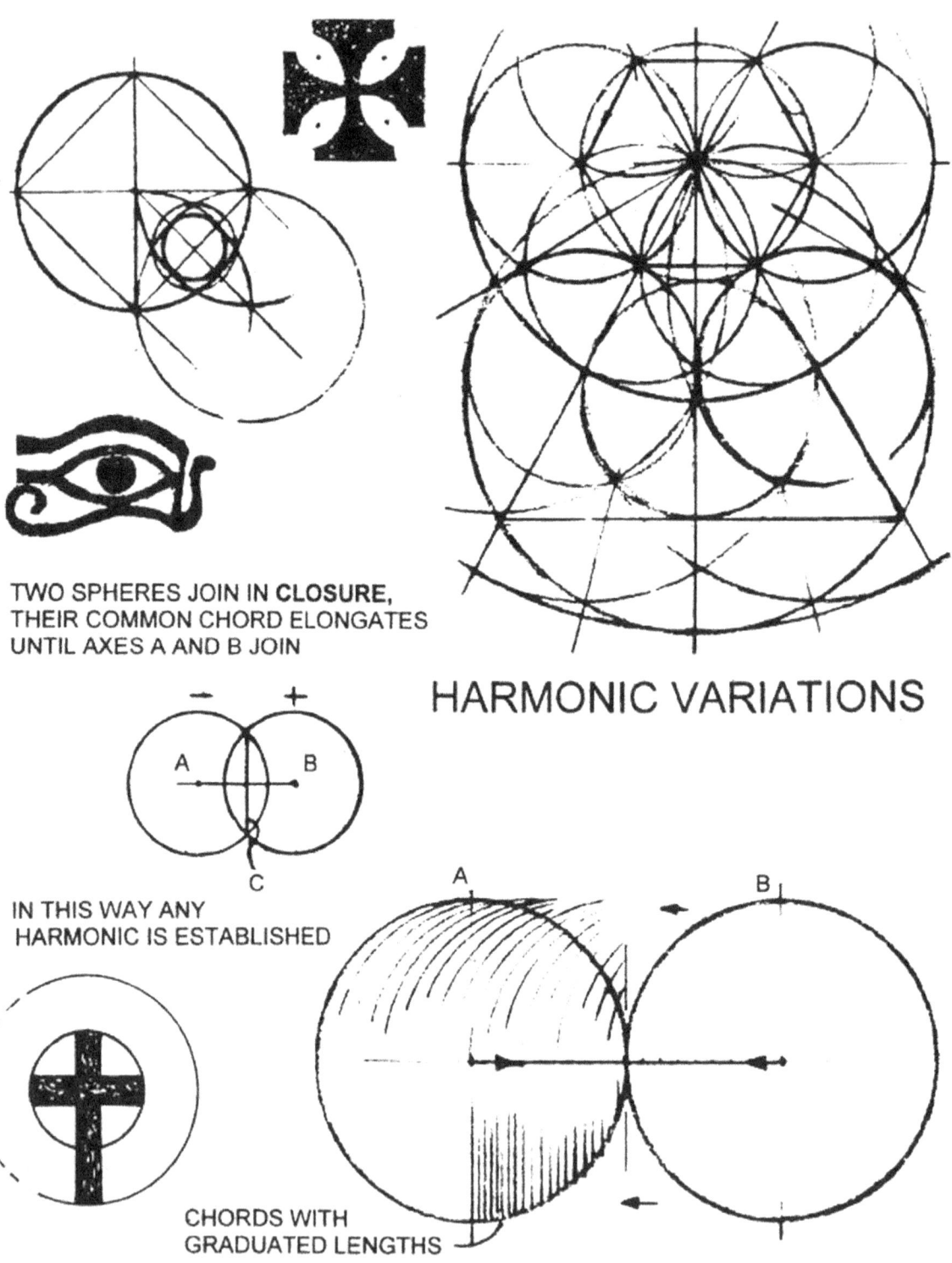

UNIVERSAL FIELDS ALWAYS OVERLAY THE ST FIELDS

As a pair of ST Fields corpuscles come together at a tangent point or at a plane, they share a fulcrum of balance at the common axis. That balance is a result of interlocking Universal Fields.

The greater octaves, the Universal octaves, prefer to exercise their greater influence along the plane of tangencies. These greater octaves combine to effect the energy and force exercised by the ST Fields. We call these fields UNIVERSAL SPACEFIELD and UNIVERSAL TIMEFIELD. Together we will designate them as USUT Fields. US FIELD = UT FIELD.

The Universal Fields have a stabilizing effect upon all galactic activity. Yet they double any force, action, or signal engendered by ourselves or our gravitational fields. The Universal Fields electromagnetic speeds double that which we know. Their wavelengths can be twice as long. Their amplitude twice as great.

> **The Universal fields stabilize the sun and the earth and all galactic environments. The USUT Fields act as the balancing fulcrum on the tangency between the S Field and the T Field and within their predominant expressions of force.**

The USUT Field has a radius double to any of the group of smaller octaves in the ST Fields. An action in the USUT Field is transferred to the ST Field as an undertone. The transfer is not quite simultaneous. The response, one field to the other, is resonant, activating the laws of each octave group. Thus the axiom "As above, so below; As below so above" is a truism. But the language and expression in signaled symbols is according to the originating creator. Translations of one octave to another can be easily misunderstood. Is there a single universal language? The only vibration that bridges all octaves is that of unqualified love. That vibration can be used as a carrier wave.

The USUT Field has larger spherical diameters, more kinetic power, smaller, slower centers with higher potentials. One field easily passes through the other, unless it elects to lock into gear at an interface.

The actual union between the ST Field and the Universal Fields is not fully understood, but it can be called an alliance of harmonics.

The resonance of tones on a piano is an excellent reference. Each octave of notes on a piano has the same harmonic intervals except to be higher in frequency or lower in frequency. When a middle C note is firmly struck, then the string is dampened, you will hear a resonant tone on the string of the C note of the next higher octave. From the top of the keyboard to the bottom, any corresponding note is reduced by one half in frequency. Your ear can easily perceive its sameness and its difference.

Now imagine that there is a piano twice as big as your Steinway to represent the Universal Fields. The overtones of its middle C note are out of your hearing range but when played your smaller piano will lock into the vibration and you will faintly hear it sound.

The body of our galaxy is our biggest overtone. The sun has the first overtone you can identify with while the tones of earth's vibrational breath echo in your body and minds with its nourishing lullaby. All life is stabilized by the song of the blue planet.

Powerful potentialities move slowly and display qualities you associate with "mass". Inertia and Momentum are the direct result of interlocking Universal Fields. This will be discussed in our chapter on gravitational forces, (Chapter 3). Needless to say, Universal Field command the forces that design the cosmology of the stars.

Within the religious practices of Native Americans can be seen remnants of history recalling ancient times when energy was more fully understood. The Medicine Wheel ceremony pays homage to the ever-present higher spirit forces that principally travel on field axes. Their references to 0 zero, points of change, their understanding of circles of union in changing relationships shows a sophistication carried along by careful remembrances of traditions. They have not forgotten the Star People and we have not forgotten them. (see **Illustration 1-9**)

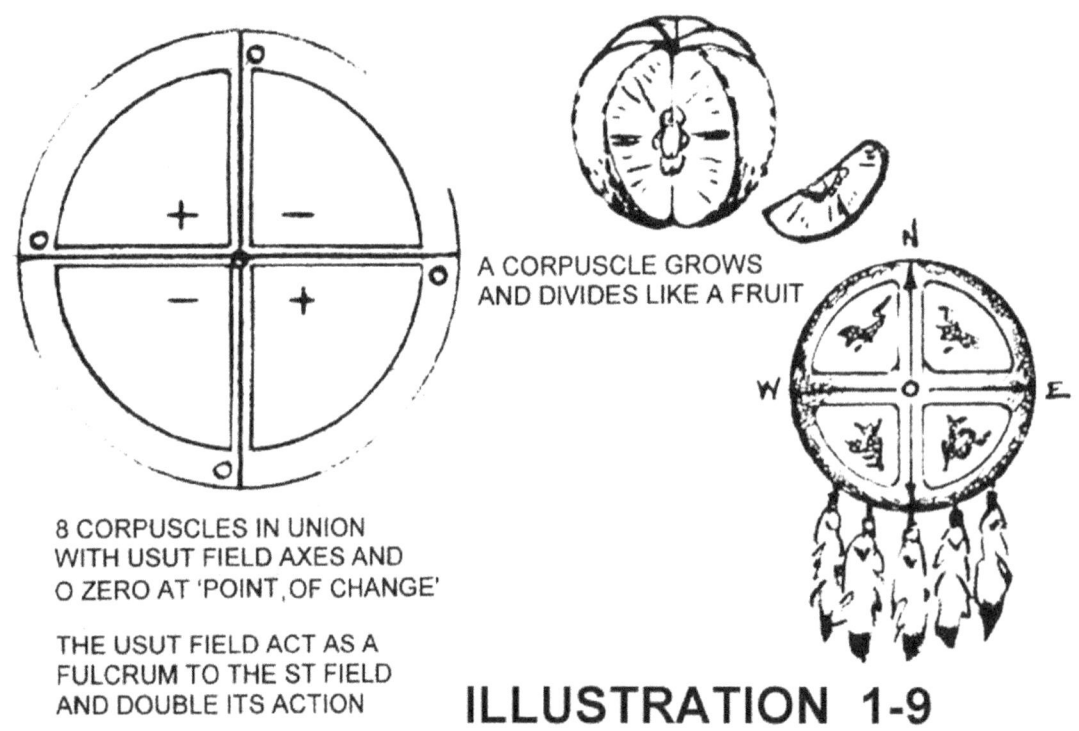

8 CORPUSCLES IN UNION WITH USUT FIELD AXES AND O ZERO AT 'POINT OF CHANGE'

THE USUT FIELD ACT AS A FULCRUM TO THE ST FIELD AND DOUBLE ITS ACTION

A CORPUSCLE GROWS AND DIVIDES LIKE A FRUIT

ILLUSTRATION 1-9

ENERGY EQUALS SUBSTANCE

Space and time are the two components of action. Both space and time are evident and measurable, yet both are invisible to a human in their abstract condition. When the SPACE Field and the TIME Field are bound together in configuration they are surrounded by a unifying sphere with a strong circumferential boundary. This spherical boundary has "surface tension" made up of tightly compacted ST Field corpuscles. This barrier boundary defines what we call substance.

By its nature S Field cannot penetrate T field and visa versa. The two fields mix in many conformations but do not blend. Each field defines the other and holds it in place. T Field has no basic dimensions, but it has place, specific locations measured by S Field. T Field expresses as gravitational radii from a "point of change". It expresses on a circumference as "surface tension". Beneath the boundary of "surface tension" a network of attractive T Field radii CLOSE inward to center. Each T Field union is resisted by the attendant equal and opposite S Field. The result is an agreement we call substance or material.

Every material is known for its "half life". The internal pressure of the union of material also causes some quanta of energy to escape the bonds at a regular rate. The continual force of contractive CLOSURE causes material breakdown with such regularity it can be used as a clock, a time unit measurement.

The union of ST Fields is the act of creation. Unions usually occur in a condition of fire. That is, harmonic speeds interlock the ST Field in very specific conformations and speeds. A variety of Universal circuits are also called fires and are specifically applied in the designs of creation events authorized by the Divine ONE.

SIGNALS TRANSVERSE THE SPACE -TIME FIELDS

> **Any action in one S Field predominant corpuscle will be reflected, equally and opposite , in the adjacent T Field predominant corpuscle.**

When alternating S Field and T Field predominant corpuscles are packed into a field like a checkerboard the field becomes a continuum that fills all space, (there is no empty space.) Because the corpuscles alternate, an action in one corpuscle, reflecting to its partner within a continuum, will reflect over great distances. This is called an energy "cascade effect". The exact signaled essence of an event in one corpuscle will be reflected everywhere.

Reflecting signals are around about and within you every day, but you may not have noticed. You are used to the idea of tossing a ball from here to there. You know it is the same ball here and there. With signals, a unit we may choose to call light may signal an event in one place and may be witnessed in another, yet the unit of light has not traveled. A signal facsimile may have been reflected throughout the field continuum. What has arrived seems to be the same light event. We can recognize our own reflection in a mirror and we call it an illusion. It is not a complete replication and it is easy to know it is not "real". Reality is not always easy to know . A virtual event can be so complete that it seems real.

Signals create illusions. Signals can create substantial illusion. Using the pure energy of the ST Field, appropriate harmonic signals configurations master-mind the creation of new energy in the likeness of the old. A single photon may generate a light signal, (A), on the sun. When that signal meets its material destination it can be measured once again as a photon (B). Photon A and B seem to be the same. You presume event A has traveled to destination event B. That is not the case. There are two photons in different places. Energy is created and transformed every second. Then what is illusion and what is not? Since signals do not travel through space, may we say that signals transit across space. Will that convey the idea?

Signals are designed and propagated by the energy of INTELLIGENCE which has no substantiality of its own. It is CAUSE. It activates its own creations. MIND is in charge. Fortunately, MIND is merciful on its grand scale as well as being inter-communicative.

An insubstantial signal has more "reality" than the fields it manipulates. The magicians trick is more "real" than the magician himself. In upcoming chapters we will try to untangle the myths of your reality as manifested through signals.

This signal transference can be as simple as light or it can compose itself into other vibrational forms; sounds, ideas, forces, materials and life energies, etc.. Thus the ST Fields become a very creative medium as it is the building block of the designers plan.

The sun shines light upon you every day, but most of this life -giving substance is not in the form of photons being hurled from the hot surface of the sun across miles and alighting upon your eyes, your grass and trees. Most of the events of sunlight never leave the sun but are accurately replicated through the spatial continuum and reactivated within your being, and within the photo-sensitive chemistry of your plants.

If you compare the sun to the likeness of a bonfire, as primitives did, then you might believe that the sun will burn up its fuel and soon disappear. You will come to learn the sun is not giving up all its energy to outer-space and that it has its own continuous fuel supply.

Signals traversing fields cause corpuscles to contract and expand in a characteristic way. This condition allows you to measure electromagnetic waves according to wavelengths and amplitude. This study is vast and will be reviewed in upcoming chapters, (see **Illustration 1-10**).

Wave signals are propagated by the essence of time-space from a center point. A signal is programmed with speed, direction and cyclic time, (as well as innumerable other details). Signals code the potential centers of ST Fields corpuscles causing overtoning and geometric shifts. Those shifting centers and their tori determine your measurements of wave form frequency. (see **Illustration 1-11**)

As electronic communication involves your lives, a happier understanding of your environment will come about. Signals become complicated between the clocking of different octaves. Think about simultaneity between undertones and overtones, between overlaid fields. That will lead you to perceive the possibilities of knowing about an event before it occurs in your earthly time frame. We will attempt to integrate what you know and what you sense. As you have suspected, substantiality is a result and not a cause, and your experiences are about to take a quantum leap into the substantial nature of CAUSE.

ILLUSTRATION 1-10

A SIGNAL TRANSITS AN ST FIELD
CAUSING EXPANSION AND CONTRACTION

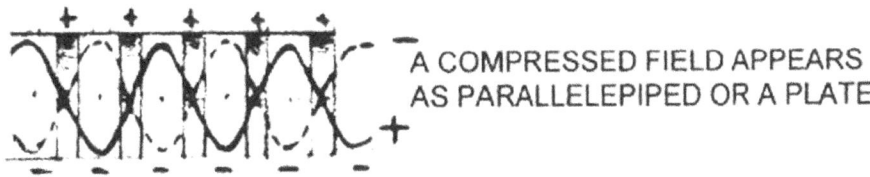

A COMPRESSED FIELD APPEARS
AS PARALLELEPIPED OR A PLATE

ILLUSTRATION 1-11

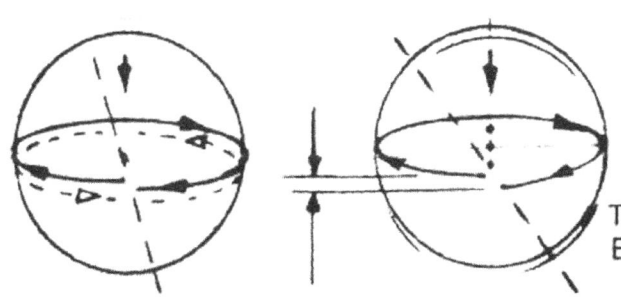

THE GEOMETRY OF TIME IS MEASURED
BY DIMENSIONAL SHIFTS AT CENTER

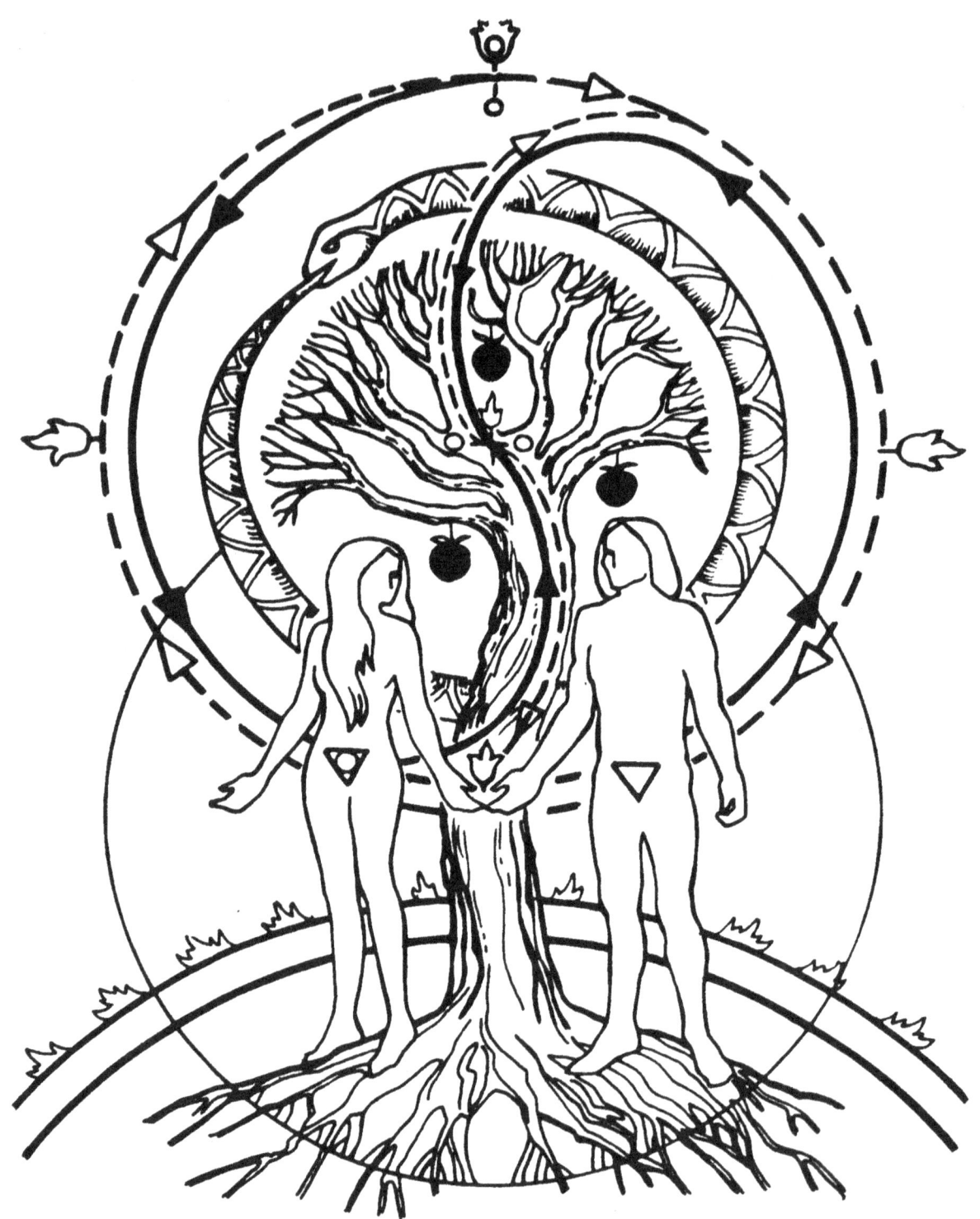

ILLUSTRATION 1-12
A NEW MILLENNIUM - A NEW BIRTHING

Biologists have presented you with some startling evidence that living creatures change themselves dramatically to ensure their survival in a particular environment. Life is self-adapting. An insect in the Amazon playas has wings that look exactly like the flowers of the tree upon which it thrives. A bird has a beak that is shaped exactly right to gather nectar deep within a particular flower in its forest. A beetle puffs itself up to double its size to frighten off predators. Most plants, animals and insects have unique biological anomalies that ensure its life cycle in a very special place. How did the creature know how to change itself so accurately? Can you suppose that a moth is smarter than you; that it can alter its own DNA?

The butterfly, whose wings look like flowers, has eyes and a sense of logic, and certainly a desire to stay alive being a butterfly. Its simple desire went forth as a signal to alert a greater intelligence who, by law, was able to assist the DNA change in exactly the way the butterfly had described. The butterfly literally said to itself, in butterfly language, "If I were to look exactly like these flowers where I spend my happy hours feeding, the predators would not see me here and I could avoid a horrible death." And so it was. Desire propels a signal to where it needs to go.

Humans use the same adaptive biological process, but they are often slowed down by traditions and lack of trust. If a human wants to fly he settles for an airplane - because he has learned to understand and trust airplanes. It is possible that a human could grow wings and fly. And then again, he wants to remain a "human". He is literally afraid of radical change from what his ideal "humanness" is like. The butterfly wanted to remain a butterfly, but one that looked like a flower. The human is very limited by his ideals. A human could live 300 years, but then he would look funny and old and everyone would hate him - he thinks.

We would like to propose that the joys of humanness could be presented in an attractive but radically different manner. If you could know yourself as a Creator Wave Form that kept your personality and identity intact, you might find it quite interesting to change forms, to alter your appearance at will. Think about what opportunities might be in the cosmos to accommodate your desires. Owls and alligators remain unchanged over millions of years. Why? Because they were so successful and happy being owls and alligators that they stayed that way. Dissatisfaction has many advantages. This is not a joke. Think about how you might want to change. Believe that you can.

UNITY IS A ROOT CAUSE

The Laws of Unity include the Law of Attraction of the physical energies. Unity implies synthesis, synchronicity, involvement, enrapture and centralization. Unity can be likened to a beloved symphony played by one hundred fine musicians and enjoyed by an audience of millions. All things of beauty come to life together in glory yet remain wholly unique.

The work that the TIME-SPACE FIELDS provides is a structure whereby interrelationships of all energies can come together in an orderly way. Order and recognition are essential in our vast Universal tapestry of life. We need not imagine that God's schemes are vague or haphazard. Even so, for all of us, there remain great mysteries and unpredictability. The mysteries interest us and we are urged on by our sense of wonder.

The T Field mechanically attracts itself. Yet the absolute inseparability of the expanding S Field from the T Field can be called a **symmetric unity**. The two opposing fields act together as clay for all cocreators with God.

In the center of unified action is stillness, oneness, the essence of unity. The truth of Unity is so powerful as to allow a creative manipulation within all octaves, energy states and coalitions. Forces are allowed to shift and flow in infinite ways.

Is it possible that anything or anyone can separate itself from the Truth of Unity? We think not. Can separation express in a polar : anti-polar condition? Yes, but not without purpose (that which unifies). Is chaos possible? Not so, if chaos means no unity at all. Purpose and Cause are the vocabulary of the Truth Of Unity, as are the words Love, Wonder, Beauty, as are the words Disarrangement and Disaffiliation.

Separation from Unity does not exist from the standpoint of the Root Cause. Separation exists only in terms of very specific equations. Imbalance stimulates the forces of balance to justify those forces. The dual fields are forever justified on the fulcrum of the Truth of Unity.

CHAPTER 2

SYMMETRY IN MAGNETISM

AN ATTRACTIVE MINERAL

When early man picked up a little rock called a lodestone he noticed it had a peculiar attraction to other little rocks of its kind. It was a day that changed the history of the evolutionary cycles of man. The iron ore in the rock acted in a fascinating way. How could a rock or a bar of iron metal forcefully attract another without ever touching it ? We shall go back to the beginning and explain what early man was incapable of understanding, and how an iron bar interacts with cosmic forces.

Like all else, an iron bar magnet is surrounded by, and immersed in a SPACE-TIME FIELD.

An iron bar is not as solid as it seems. The atoms inside the metal form long lines from one end of the magnet to the other. The atoms oscillate. In between the atomic lines there is a ST Field. Inside a bar magnet the atoms act like propellers on a motor boat. They push the ST Field out one end of the magnet and pull it in the other. The ST Field flows through the magnet at the rate of the atomic thrust which is slower than the flow of gravity.

You will remember that S Field moves equal and opposite to T Field. The shape and spin of the atoms in the iron bar magnet determine the fields directional preference . The S Field is driven one way, and the T Field is driven the opposite way, consistent to the atomic alignment. It can be demonstrated that the T Field is driven out the N (north) end of the magnet and the S Field is driven out the S (south) end of the magnet. The number of atomic lines and the density strength of the magnet will determine how far out, away from the magnet, the surrounding ST Fields are effected by the flow. The number of atomic lines equals the number of flux loops that circulate around the magnet.

You will remember that it is the nature of T Field to CLOSE in on itself and it will "take its own tail in its mouth". The T Field has been given a new speed and vibration by the atomic spin of the bar magnet. The new speed means a new energy octave. The T Field, emerging from the magnet at the N end, seeks to join with itself – of that speed. It searches and finds its own beginnings and likeness near the S end of the magnet. The T Field begins a circuit loop through the magnet, around the outside, then back through the magnet. Each passage through one atomic line may form a separate circuit. A magnetic flux loop demonstrates the very nature of T Field.

You will remember that the S Field, moving at the same speed as the T Field, but moving in the opposite direction, circles and forms a swirling, spreading sheath around each T Field circuit line, giving it definition and separating one T Field line from another. Bundles of lines form and spread outward in an apple shape around the magnet. Equal pressures distribute the lines evenly, layer upon layer.

We are taking you step by step through this field activity of a magnet so that you will begin to understand how the LAWS given in Chapter 1 are applied in your experience. **Illustrations 2-1, and 2-2**, will make these ideas graphic.

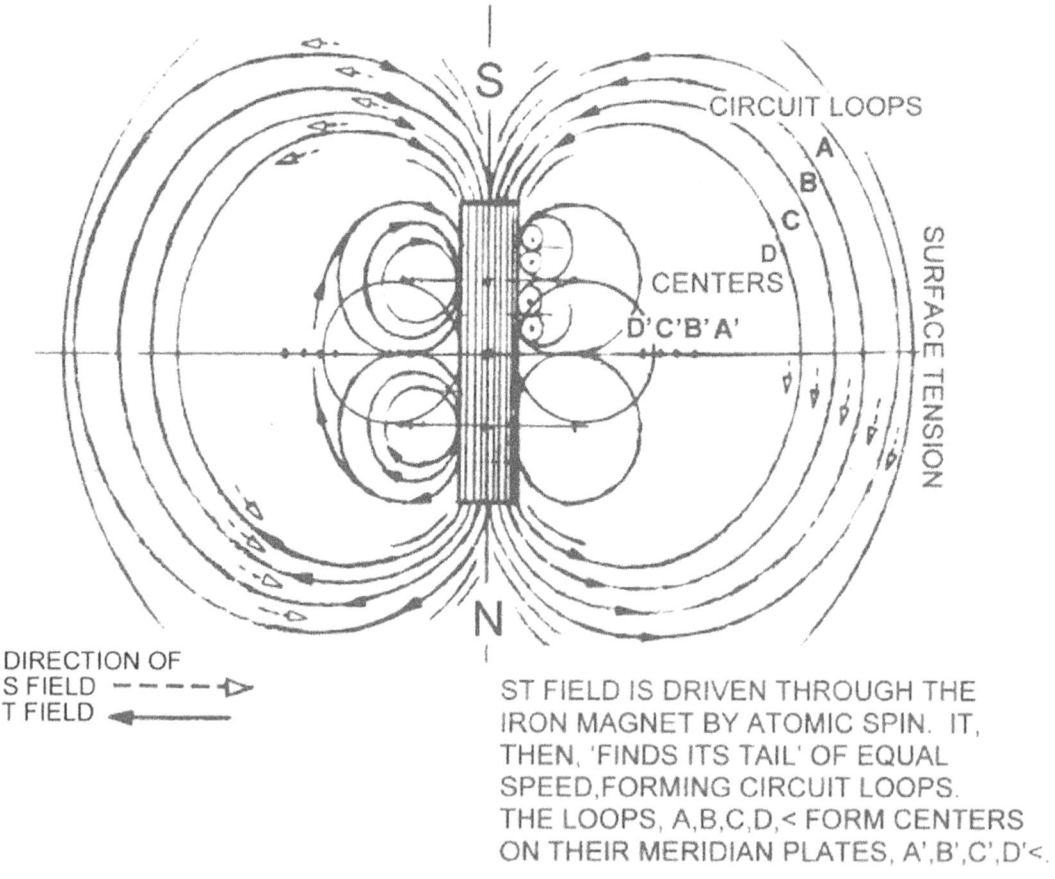

This is your first study experience with directional field force. Motion in a field activates force. **Force is, however, only functional when it confronts something "like-unlike" or equal and opposite itself. Force acts upon fields of its own octave speeds and is not noticeable as force in other octaves. The nature of force is that it is unique to its own octave.** This fact is vitally important to all engineers.

When a second magnet is placed N end to N end of the first, a butting of directional forces prevents a union. When a second magnet is placed S end to S end of the first, a butting of directional forces prevents a union. Force flows are pushed aside to prevent collision. The S Fields when placed in adjacency will not attract each other, whereas T Fields will normally CLOSE when adjacent. However, opposing flows of fields disallows a union, even in T Field forces.

The sense of **repelling** of magnets comes from active field flow forces coming from opposite directions, within the same octave.

When a second magnet is placed N end to S end of the first, the lines of force are fully compatible. Then the T Fields will CLOSE in a normal way. S Fields continue to wrap around each T Field line. One T Field circuit is formed instead of two. A field sphere encompasses the whole. (see **Illustrations 2-1, 2-2, 2-4a**)

You will remember that any resonant action in an ST Field develops a sphere with a center. A T Field center will form in the middle of any length of bar magnet. Its sphere will extend to the ends

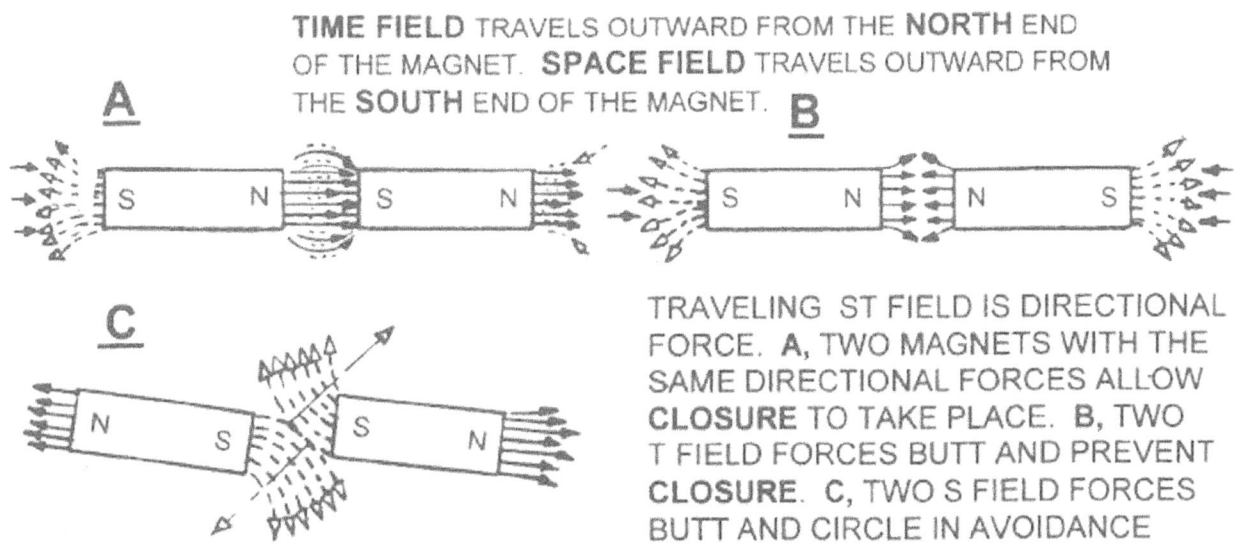

of the bar, making the bar act as an axes. A sphere will extend beyond to encompass the whole of the magnet and its field circuitry. All axes will develop and divide as in all natural field divisions. The secondary Y/ Z axis becomes prominent.

Each line of flux that travels through the magnet and around the outside in a loop is, in fact, a flattened sphere, which looks like a plate. It is a polarized sphere. The sphere is so deformed by pressures around it that it becomes almost two dimensional rather than three dimensional . The center of this flattened sphere lies on the equatorial Y/Z axis. We can call the flattened sphere a meridian plate circuit. Since each line of atoms creates one meridian plate circuit there will be a million and more of these plates packed around the magnet in an apple shape. S Field surrounds and pushes outward from each circuit, dividing each plate one from another. Each distinct meridian plate circuit will have a distinct center point (of appropriate quanta) all laying upon the equatorial plane of the magnet.

The ST Fields being driven through the atoms of the magnet appear as a forceful wind at either end of the magnet. It is important to notice that the normal union of these forces takes time. The process is particularly slow. A small magnet, if rapidly passed perpendicular to the force field of a larger magnet, will show little attraction. If moved very slowly through the force field an attraction is evident. CLOSURE will take place as lines of force harmonize. The whole of each sphere has to sense the harmonies of the other and agree to form a common sphere and axis. Harmonic integration is essential for radii union and CLOSURE. This phenomena plays a vital role in engineering.

When any ST Field flows like a river, or a wind, it exerts force. That force is only effective within its own octave ranges . You will notice that an iron magnet does not exert a force on a bar of copper or brass. The atomic structures differ. The metals do not have compatible vibrations.

Notice that the lines of a moving field known as gravity exert force on iron, copper, brass, cardboard, and all other materials on earth. Gravity is a grand overtone of oscillations and engages all. Gravity will be discussed in Chapter 3, in as much detail as our overview of physics will allow.

SEEK AND FIND

It is a phenomena that a T Field line can recognize its own kind over a great distance and then proceed to CLOSE with it. No moving field line is traveling alone. It is always accompanied by the USUT Fields which have great outreach.

As a field line bifurcates, forms a center, it also describes the largest possible sphere around the whole. This sphere, like all others, has "surface tension " which is both S and T Field in motion. Within and around the whole sphere there is communication along its radii. The magnetic python can easily seek and find its own tail in this circumscribed environment. As soon as a potential center is formed, an exact radius defines its widest circumference. "What reaches out, comes around."

Because any length of magnetic field line continues to bifurcate, the spheres continue to form relative to the radii formed by bifurcation. And because of these divisions, the whole sphere develops concentric ringed patterns (see **Illustration 2-1**). The packed meridian plate circuits follow these concentric paths. (see **Illustration 2-3**).

It is very important that you visualize the invisible sphere and circuits around magnets. The active principles in these fields apply in many more ways, as you will soon see.

ILLUSTRATION 2-3

THE COSMIC - ELECTRIC PHENOMENA IN MAGNETS

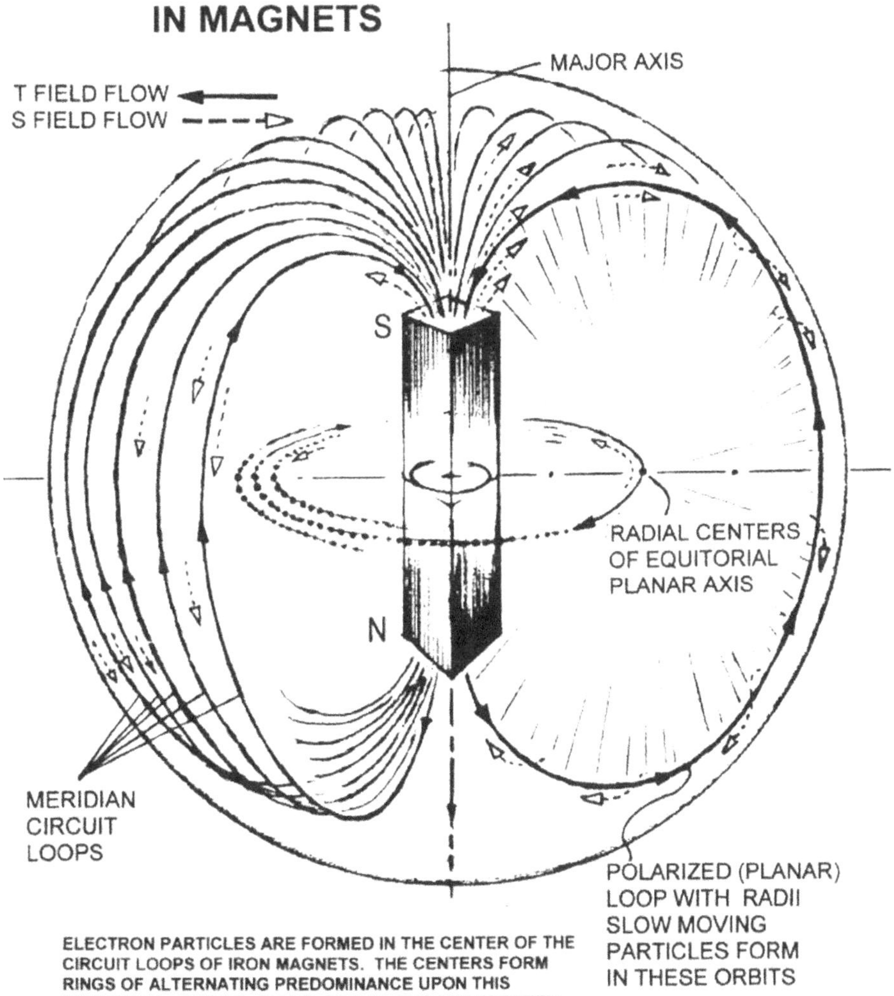

T FIELD FLOW ←
S FIELD FLOW ◁---

- MAJOR AXIS
- RADIAL CENTERS OF EQUITORIAL PLANAR AXIS
- MERIDIAN CIRCUIT LOOPS
- POLARIZED (PLANAR) LOOP WITH RADII SLOW MOVING PARTICLES FORM IN THESE ORBITS

ELECTRON PARTICLES ARE FORMED IN THE CENTER OF THE CIRCUIT LOOPS OF IRON MAGNETS. THE CENTERS FORM RINGS OF ALTERNATING PREDOMINANCE UPON THIS EQUITORIAL, ORBITAL PLANE. ELECTRONS ARE CREATED FROM THE FIELD AT LARGE UPON THESE SLOW MOVING ORBITAL RINGS. WHEN THE RINGS SHIFT POSITION THE ELECTRONS ARE RELEASED.

Copyright 2001, A.V. Olson

WHAT IS ELECTRICITY ?

Magnetic field forms around the outside of a magnet. This field consists of meridian plate circuits compressed and layered into an apple-shaped fan around a magnet. Each of millions of plates alternate with T Field predominance and S Field predominance. Each has a center which lies on its appropriate circle upon the equatorial plane of the magnet (see **Illustration 2-3**).

We have said that each T Field circuit loop (the snake biting its own tail) joins itself and re-enters the magnet. Now here is a new scenario. Each T Field line loops around and grabs hold of the tail of its neighbor line. The neighbor line has no choice but to grab the tail of its neighbor, and so on. Within a second you have continuous looping lines called a toroid. Inside the magnet, the atomic motors continue to drive the field flow in one direction. The flow of field in the toroid now has a secondary direction perpendicular to the first. It flows around the apple, spinning the equatorial plane like a phonograph record. The direction of flow is always the same, defined by the galactic toroid. Current directions are best graphically described by Ampere's right hand rule, as shown in **Illustration 2-7**.

Whether the circuit loops are acting individually, or if they are acting together as one continuous circuit, each loop has a center which lies on a common plane.

> **Each T Field center of a T Field circuit loop has a potential relative to the speed of the circuit flow and according to its own speed of travel. It carries a discrete quanta of energy.**

Each T Field center is discrete, surrounded by S Field predominant centers and sheaths. Each quantified center point is exactly held in place by the positions of radii of the circuit loops.

> **Each T Field center on the XY axes has been created from ST Field around the magnet proportional to the field flow of its circuit loop.**

Each center spins with other discrete centers on an orbital path. Because the circuit loops are bundled and layered in orderly concentricity, the centers form concentric orbits upon the equatorial plane. Because of bifurcation there are a series of equatorial planes focalizing along the length of the magnetic bar. **Illustration 2-3** can help you visualize the circuit loops and centers. To avoid confusion many layers of loops and planes are left out of the drawing. **Illustration 2-1** shows loops and planes from bifurcation.

Here we are at your magic moment!

> **When the field flow in the circuit loop makes a radical shift, the center (previously created quanta of field) is set free of its location.**

This center, which you call an electron, will travel on any T Field path to 'ground'. It is now a *free electron* which can be carried off as a slave by a simple copper wire. The new position of the circuit loop forms a new center. Once again, as the circuit makes a change, an electron center is set free. This is how you power your civilization: courtesy of the ST Fields, and a few magnets.

These dynamic centers are, of course, traveling slower than their parent circuit loops. They have exactly the potentiality that you can make use of.

If you project your imagination to include other metals, other crystals, you can see that there are other quanta that can be developed from the ST Fields that may have other purposes than the generation of electricity. You can also guess that you can, at will, alter the speed of field flow. If you replace a magnet with a quartz crystal wand, imagine its field loops with centers, what do you see in its central orbits? How does it feel?

ELECTRONICS

The electron centers move when (1.), The circuit loops rise or fall, or (2.) , The circuit loops show a steady flow. You know that by alternating the poles of an electro-magnet the circuit loops show sudden falls to zero, thus shocking the center out of position and onto a grounded wire. As an old center is taken away a new one forms in its place. By this means, you create electricity. You create usable energy directly from the ST Fields.

Conversely, if you wind a wire into a coil and make it part of your circuit (a solenoid), you can duplicate a magnetic field flow in something other than an iron magnet. Unnatural magnets are made in this way. (see **Illustration 2-7**).

A current traveling around in a toroid creates a dipole. A dipole is a vortex. The flow of field within a dipole loop is of great interest to us (see **Illustration 2-4**).

You have developed a vast knowledge of how to use an electron. Your electrons have a defined potential (speed). You also recognize that both S Field and T Field interact in discretely charged particles. The idea that can be demonstrated by a magnet is the same ideal pattern that expresses in an atom, a molecule, a charged particle, a spark, etc. A photon has an S Field predominance, whereas an electron has a T Field predominance. Any center, created and released, creates a signal carrying its own characteristics and vibrational imprint. The signal radiates out into the fields telling the story of the event.

An electron activated in a vacuum has the advantage of working in an area of pure field without the interference of oxygen and other gases. (ST Field is not confined by any material, such as a bell jar, but materials will slow down the passage of fields. One of the reasons you know that magnetism is a pure field flow is that you cannot put a magnet any place where it does not function). An electron will combine with gas molecules all too readily. So you have made use of the many tricks an electron, or an electron signal, can do.

In the process you have ignored your production of S Field predominant units. What can S Field particles do for you, as they travel through your space? S Field does not stand alone. If partners are not available they will create partners from pure field. Magnetic rings circle around currents. Widespread toroid circuits develop around generators. The speed that identifies the electron

ILLUSTRATION 2-4

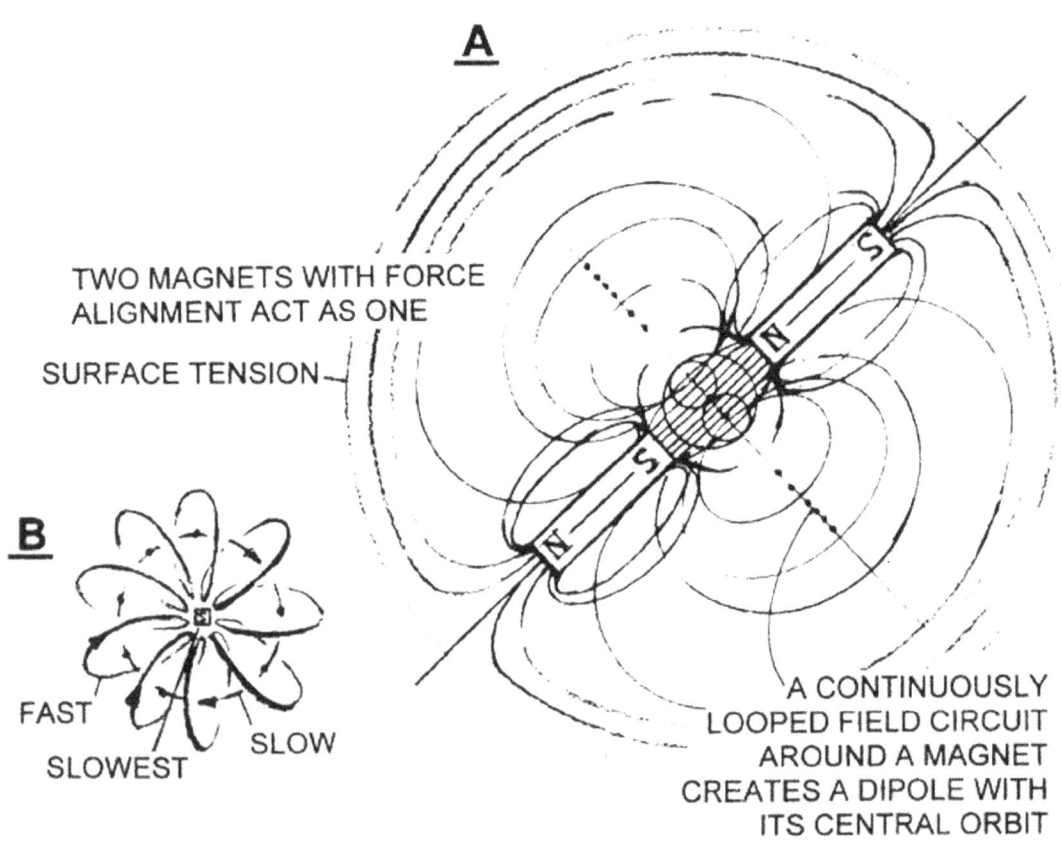

A — TWO MAGNETS WITH FORCE ALIGNMENT ACT AS ONE
SURFACE TENSION
A CONTINUOUSLY LOOPED FIELD CIRCUIT AROUND A MAGNET CREATES A DIPOLE WITH ITS CENTRAL ORBIT

B — FAST / SLOW / SLOWEST

unit and its partner in the S Field is not the best for animal cells. That speed (quanta) throws the balance of fields too far into the heavy metal octaves. The human spirit seeks a much finer, higher, faster frequency. The human soul finds that even the speed of gravity is burdensome. An animal body needs to soar higher to maintain proper growth and health. This is a serious matter to consider because your environment is so crisscrossed with electron signals and particles. Field signals at the electron level are not easily blocked out.

It places a burden upon the individual to alter the signals within his/her own body, certainly within the brain and chakras.

How can you recognize a finer vibration? Sense the vibrations of certain crystals. Around every stone there is an individual field which you may learn to identify. By sensing the vibrations of plants, their flowers, and seeds, with your sensory facilities you can learn how to keep your own vibrations flexible. For instance, you can learn to identify the smell and vibration of a clean lepedolite stone (which will relax you). Then later remember that vibration and imagine that you are that stone. You can learn to identify a wild rose, see its color, smell its fragrance, taste its petals (asking permission) and take in its total vibrational pattern (do not pick the rose). Then later remember those experiences fully and pretend you are totally that rose. Then expand that signal outward to thank and honor all wild roses. This exercises your own god given "electronic" systems, but at a life-producing frequency level. Believe that you can do it fully. Practice.

Quartz stones, tourmalines, beryls, spodumenes, celestites, lepedolites, and many more stones emit valuable vibrations that your body can detect. Each has a specific correspondence to your biology.

Colors, used in your personal way, can modify your emotions and bodily conditions. For instance if you find yourself angry at someone, try imagining a blue-violet glow in the area of your thyroid. Then place the pink glow of love at your diaphragm. Focus on that and do not take on the vibration of anger. Expand a pale blue glow around your head, then around the whole body of your protagonist. You will not re-act foolishly.

Is this a scientific approach? It certainly is.

Let us bring the dipole to mind (over and over again). If a loop of field circuitry, traveling at a specific speed, encloses an area of field, something happens in that field. (see **Illustration 2-5**). Bundles of fields begin to move and flow like a river through the vortex in the direction of the galactic principle of the "right hand rule". (see **Illustration 2-7**) In addition, the entire speed and vibration and detailed signal is copied in all parts and all bundles of the field now moving through the vortex.

> **Field flow through the vortex of a dipole carries the complete message of the dipole current in every part of the orifice. Any identifying condition of the toroid dipole is multiplied any number of times within the orifice of the dipole.**

Here you see the principle of **S Field multiplication** at work. Your system of mathematics is validated by cosmic LAW.

An imaginative student of science will begin to see the unending applications of this law. Soon you will see how to demonstrate the truth of this law.

For now, believe that it is true, and reverse the idea. Any field flow through a vortex creates a current in the dipole which duplicates accurately the messages of the flow. You now have a new idea introduced, that current or field flow carries detailed messages with vast amounts of encoded information. How do we know that?

You experience with your eyes the coded information of light, and your eyes are designed to decode that information. By placing a pinpoint vortex within a larger vortex one set of codes can be isolated. The pupil of your eyes selects one bundle of coded light.

An harmonic wavelength travels to the plane of retina in your eye to "resolve" the details of the coding. The retina converts the signals into discrete quanta units which travel as current to the brain where it is interpreted as a picture of light. What a marvelous "electronic" mechanism the eye is!

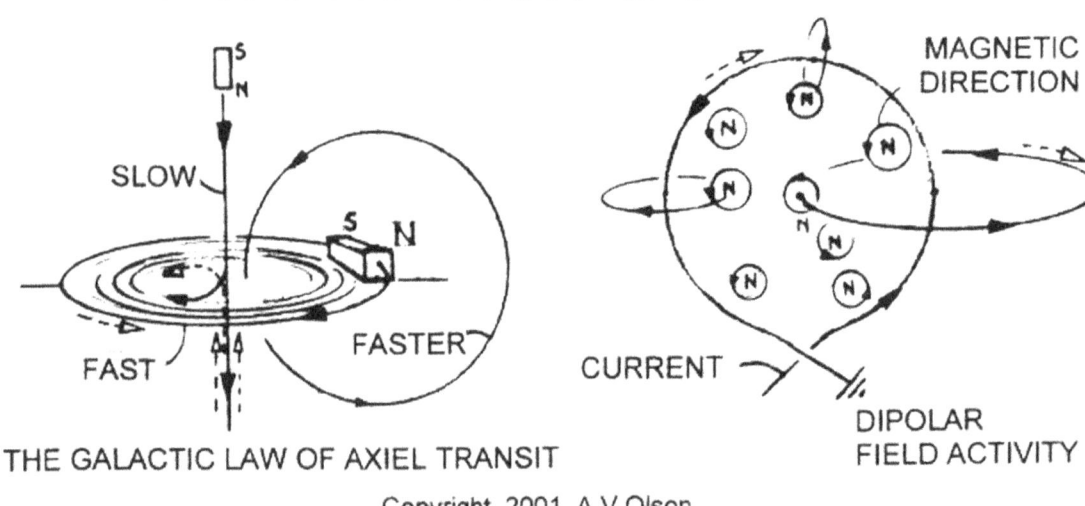

ILLUSTRATION 2-5

THE GALACTIC LAW OF AXIEL TRANSIT

DIPOLAR FIELD ACTIVITY

Copyright 2001, A.V. Olson

Your body and brain are full of eyes. One of the most useful is the "third eye", an invisible orifice in your forehead, above your nose. Believe that you have such an eye, even though it has not been found in dissection, a current loop dilates to send and receive coded information which may be of interest to you. There is a similar orifice over your ears that decodes subtle sound. There are similar orifices at each chakra point, and each sensory organ.

The dipolar current around the third eye flows as ST Field carrying coded, signaled information, passes through it. The current is then carried into the brain. The current of the dipole can be coded by yourself and sent outward. Which way would you move the current in your third eye dipole to send a message outward? Figure it out.

If the third eye dipole has a basic vibrational tone or color it can alter the condition of the signals moving in and out. An unhappy child is apt to add the color of anger to messages moving in and out even through the adult years. Is your third eye open? What is its color this day?

The incoming field messages can be negative, or heavily laden with electronic coding, as is common in cities. The current you project into your third eye dipole can alter the quanta of the incoming field by expansion. Increase the diameter of the dipole and color it golden. Create expanding concentric rings around your head and around each vortex. Give special care to your throat chakra. It can act as an electronic 'ground' , just as the third green wire on your appliances 'grounds' excessive charges. Practice altering the speed and coloration of the field currents around your third eye. You don't need a college degree to do it.

When you sense danger your vocal chords (at the place of throat chakra) suddenly swell up tight. Messages are directed "electronically" to this place. Air is taken into the lungs and let out through this orifice as a scream. A scream is most often involuntary . Your biology was designed for its own automatic protection (like a smoke detector in your home).

Your ears are designed to detect air compression. You would have trouble hearing in a vacuum. However the "hearing" part inside your brain responds to field signals. You can hear without ears. There are devises now on your planet that can broadcast radio messages directly to your brain in full fidelity. When you hear these broadcasts, can you find the way to turn off the sound? It is very important that you do. Your life may depend upon it. On the bright side, many deaf persons will be made to hear when this devise is made available and made safe.

The sense of touch in your fingers and toes is very sensitive to field vibration. The sense is so keen it can detect and differentiate atomic oscillation rates. Practice in a quiet place with a blind-fold.

If you are in a completely darkened room and wearing a blind fold, you will be able to see your hands and fingers in front of your face as black on an almost black background. Field signals transmit into your closed eyes in quanta much finer and faster than light. Your brain can interpret these signals. Once the frequency of brain signals is understood then a different reality will become available to you. Again, you may extrapolate the dangers of possible intrusion. However you are not without biological protection if you learn about what is yours to use.

Remember that among the genes you actually see in your DNA studies are invisible genes full of circuitry. And remember from the ancient past, that genes from the Star People mingle among your own. If you turn over your biological "electronic" system to man-made mechanisms you will be slowly debilitated. Use it or lose it.

Your planet is balancing on the edge of environmental disaster. But then the human race has to do, in order to *know*, in order to *do not*. Experience, it seems, is the only teacher. Some of you catch on quicker than others, and can help others to learn about human sensibilities.

THE COSMIC CONNECTION

The spherical field design of a small magnet is very much like our cosmic arrangement of suns and planets. Let us bring the differences and similarities into focus for you, (see **Illustration 2-6**).

The earth has a geographic axis around which it rotates. All planets and suns rotate (although not at the same rate). Rotation occurs because all planets and suns have toroid circuitry, circuit loops that spiral. The loops of a small magnet do not spiral until it is designed to do so.

The earth has a separated geomagnetic axis about 12° away from the rotational axis. This axis gets strong response from nickel-iron metals. This axis shows evidence of having circuit loops just like those of a simple iron bar magnet. Iron ores all over the earth try to align their atoms with this axial pole. Copper does not. The N end of the geomagnetic axis is in Antarctica. The S end of this axis is in the Arctic. This axis generates T Field currents running around the earth's equatorial plane from east to west. It has strong field winds that help move air and water. This motion is consistent with the "right hand rule" of axial relationships. The geomagnetic axis is harmonic to the earth, but has a limited effect on materials. Massive amounts of iron deep in the earth's center are largely responsible for the strength and position of this axis. Created when the earth was molten, the axis became fixed in the metal core. It is largely cosmically independent. Because the central iron core is surrounded by molten rock it is free to shift positions in response to cosmic influences.

The third axis is the gravitational axis. The gravitational axis acts in a different way than either a magnetic axis or a rotational axis. It is very responsive to cosmos forces yet it is the vital identity of the planet earth. It lies perpendicular to the orbital plane of the moon. It holds the moon in orbit and keeps you from flying off into space. The next chapter will detail its functions.

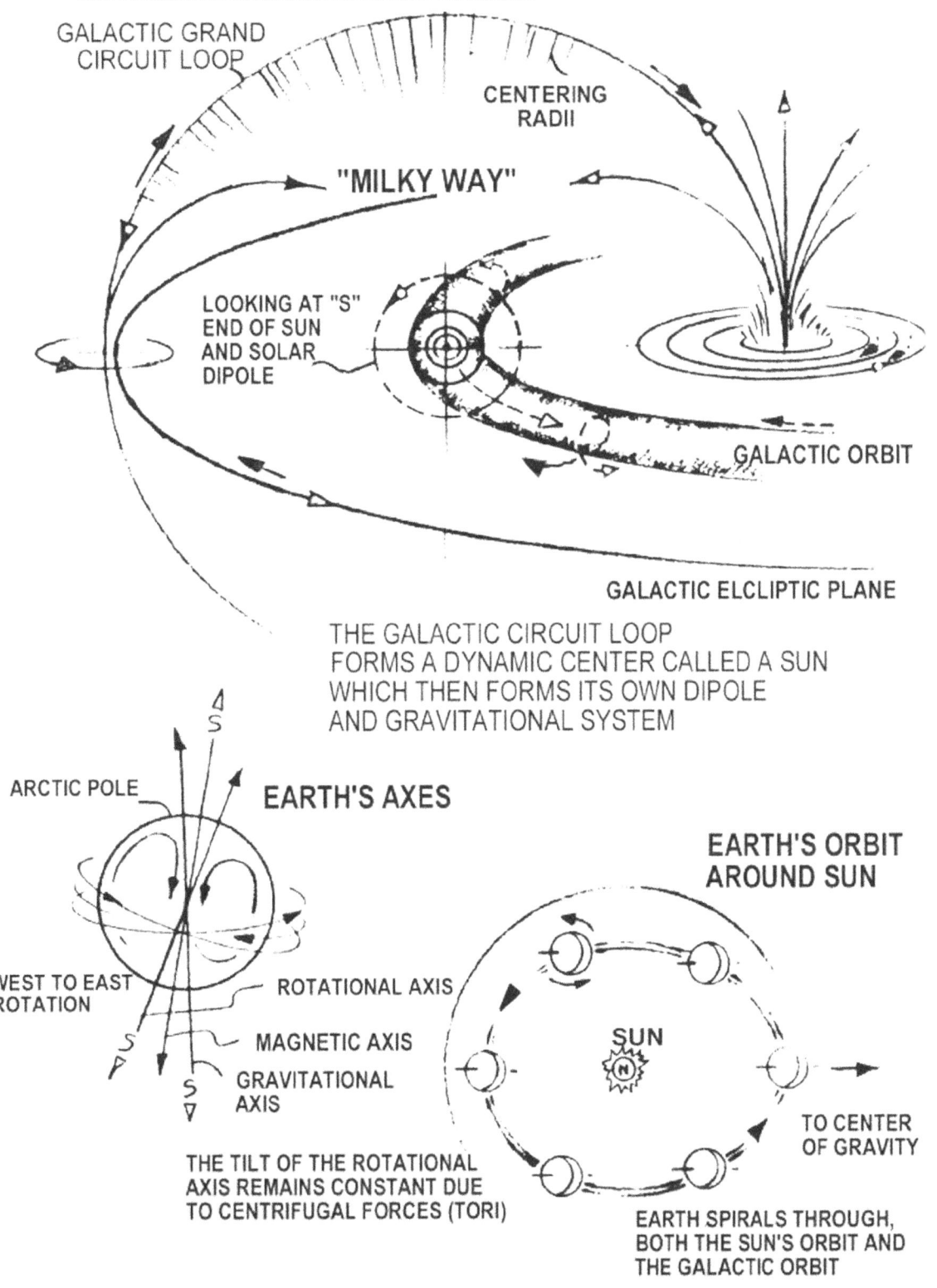

Yet all three axis together function as a dipole and were caused by evolutionary cosmic forces according to Universal Law.

The geographic axis shows no typical mineral response. It is located by watching the heavens above as stars seem to swirl around a center point. This axis is now 23° tilted to the solar ecliptic (the sun's equatorial plane). When the earth was hit by a very large comet thousands of years ago, it was tipped to one side. It continued to spin as it always had before the hit. When the earth was forming, its rotation was caused by the strength of circuit loops from the sun that spiraled through the earth's center axis. The solar axis of a planet is always perpendicular to the solar ecliptic unless cosmic bombardment throws it off, or internal shifts of material torques the spin. Material "mass " and irregular circuit loops interact to skew axes and orbits.

The planets are children of the sun. To help understand them we can make an imaginary leap to how the sun was born.

The center of the "Milky Way" galaxy is not visible to you in your octaves. It is there in all its majesty weaving the tapestry of its giant, swirling family. It extends its dipole's gravitational influence far, far, beyond its faintest outpost star. Its loops circle outward, spinning in dark power and centering quieter places where suns form, (see **Illustration 2-6**).

Let us digress for a moment to the very small scene of a laboratory where a scientist runs a current through a wire making a circuit. He finds that "magnetic" rings form around the wire, perpendicular to it. The rings are concentric and alternate in direction. They are the ST Fields doing their dipolar tricks again. He also finds that another wire, parallel and near to the first, and properly grounded, suddenly shows a current moving in the same direction as the first. The second wire has passed through the "magnetic" rings of the first, and demonstrates all the replicating aspects of a dipole.

Now imagine the first wire with its series of concentric rings around it. If you could see these rings, you might see that they join in a spiral of S Field which travels in the opposite direction to the current.

Now the scientist twists this wire into a loop. He has constructed a toroid. The spiraling concentric rings remain in their perpendicular axis to the wire. That means that the rings bundled together in the center of the loop, all traveling in the same direction in a fountain of concentrated energy. A dynamic vortex has been created (see **Illustration 2-7**).

Once again we think of our "Milky Way" galaxy with its uncountable energy circuits reaching around in an apple-shaped toroid. The circuit loops create dynamic center points, which all taken together, form orbital rings and points on a galactic ecliptic.

ILLUSTRATION 2-7

SOLENOID

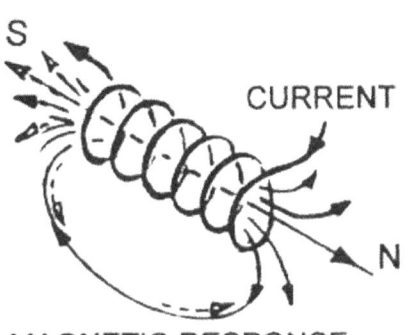

MAGNETIC RESPONSE

ST FIELD REFERENCE

THE RIGHT HAND RULE

FOR AMPERE'S LAW: THUMB POINTS WITH DIRECTION OF CURRENT. FINGERS CURL WITH DIRECTION OF MAGNETIC FIELD

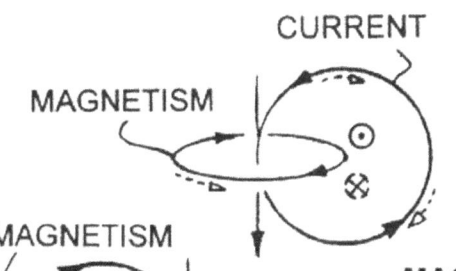

MAGNETISM AND CURRENT ARE THE SAME EXCEPT FOR DIFFERENCES IN SPEED AND FIELD PREDOMINANCES. THEY ALWAYS KEEP THE SAME AXIEL GALACTIC RELATIONSHIP.

DIPOLE VORTEX

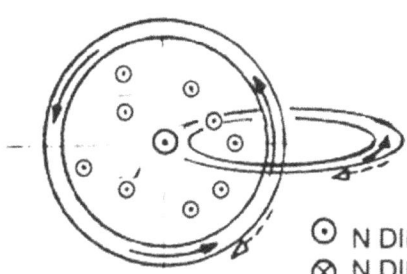

TOROID

SYMBOLS FOR NORTH AND SOUTH DIRECTIONAL CURRENTS

⊙ N DIRECTIONAL CURRENT COMING OUT OF THE PAGE
⊗ N DIRECTIONAL CURRENT GOING INTO THE PAGE

◌ S DIRECTIONAL CURRENT COMING OUT OF THE PAGE
✹ S DIRECTIONAL CURRENT GOING INTO THE PAGE

Copyright 2001, A.V. Olson

One circuit loop has a center which you know as your solar system. One circuit loop supplies the energy and identity of your sun. Once created, the sun grows in independence as it is equipped with its own gravitational axes and field circuits. It is designated by MIND with signaled attributes and personality. It is nurtured according to plan by the personalities of the galactic toroids.

Energy is seeded with a program for hydrogen and swept onto the equatorial axis to be spiraled outward onto a cooling sea of fields.

At its formation the sun was a point in fusion radiating a vast plane of hot gasses. Note that the solar equatorial plane is the same as that of the plane encircled by the circuit loop of the galaxy. That plane lies more or less perpendicular to the galactic equatorial plane. The arctic pole of the earth, matching the "N" pole of the sun, travels forward in the galactic orbit. As the solar system orbits the galaxy it passes through a series of S Field predominant bands alternating with T Field predominant bands. The regional changes in the solar environment are significant to the nature of life on earth.

The sun did not explode, it grew and spun off its veils of primary gasses as far as the edges of the toroid. Each atom of gas found companions for grouping. Each atomic group found complimentary orbits into which it could seat itself to rest. Each orbit gathered its own kind according to speeds and the paradigm of the solar circuit loops.

The sun is firmly centered and energized by a grand toroid . The sun, in turn, reaches out to its many circular arms to hold each orbit fondly in place. Each planet, resting in its orbit, reaches out to moons and particles holding the family of material worlds together.

One solar T Field circuit loop enters the top of each planet and exits out the bottom, (Arctic to Antarctic). It established the geomagnetic axis. Please refer to Illustrations 2-6 as the geometry of planes and axes can be quite a puzzle. The grand galactic dipole governs the solar ecliptic, the orbits, and planetary rotation. Your master clock and governor is your galaxy, not your sun.

> **The galactic dipolar toroid, a grand circuit loop, finds its' radial confluence at the sun, which it created and sustains.**

The sun is not a bonfire hanging in the sky, as primitive peoples believed. It will not burn itself out. It is continually being fueled by the galaxy. Our Universe is organized under LAW and very little is left to chance and accident.

CHAPTER 3

THE CREATIVE BREATH OF GRAVITY

GRAVITY FLOWS UP AND DOWN

Gravity is a unique field phenomena that functions under complex laws. Earth's gravity is hers alone, for it contains all the vibrations of all materials and life forms of this planet. It breaths in overtones the cycles and energies unique to this watery blue address in space. Gravity holds it all together. It uplifts and puts down. It translates the cosmic music into tones human beings can understand. It is a creative center which closely resembles the great cosmic center which we call the "Eye of God".

Gravity begins with a centering of a point of potential energy upon a strong central axis. The potential center of earth, provided by both the sun and the galaxy, is a seeded replica of that which centers Universal forces. Its axes are compounded to serve the activities of MIND. The WORD is spoken. Life dwells in the center and expresses to the outer edge of its sphere of influence. From the totality of the sphere it signals and receives. It performs like a self-conscious broadcasting station.

WHAT IS GRAVITY ?

When a collection of gasses and particles gather together in an appropriate orbit, they begin the process of T Field CLOSURE. These particles are of mixed fields, both S Field and T Field in atomic arrangements, but with T Field predominance. They gather in along the paths of prescribed speeds. Each tiny spherical entity joins to make larger spherical entities, until one large sphere predominates. The center of a sphere in rotation is its slowest place, and all the T Field points try to accommodate to that slowest place. In order to do that with a mixed field, a field separation must take place. S and T Field separate in a fiery disassociation. The separated T Field CLOSES to center, S Field is released to travel outward to the circumference of the total 'sphere of influence'. The force of CLOSURE of a planetary core is strong enough to pull the fields apart, causing explosive heat.

Accumulated materials around a planetary core are systematically circulated in order to internally interact with the core. The core grows larger, the earth's surface shrinks accordingly.

Separated S Field accelerates outward to the exact radial distance required by the potential of the T Field density at center. The speed at the circumference of the outermost sphere is in equal and opposite ratio to the inner central potential so that the result is ONE.

S Field circles out from center as it is released by the fusion process of the T Field core. It spirals in acceleration to the circumference, reaching a maximum speed required by the center potential then makes a mysterious and dynamic change. That place and speed can be called "point of change = 0." At this place S Field changes into T Field, slows down, and drops back to center. The spherical circumference is covered with these events of field metamorphosis. The mixed field at the circumference is said to have "surface tension". A surface tension acts like a thin shell on a sphere that pulls to center along radial paths and holds the entity of the sphere together. While defining the integrity of the sphere it can act as a protective barrier.

As the T Field forms at the circumference and begins its descent to center it decelerates at the same rate that the S Field accelerates in the opposite direction. The T Fields follow radial paths that sweep inward through the sky, picking up stray T Field particles and pulling them center. The T Field moves from the "point of change = 0", to a line and to a center point which becomes another "point of change = 0". The descending T Field lines are wrapped with ascending spirals of S Field, keeping each line apart from the next . The T Field gravitational lines pull to center, continually slowing. At a certain place, deep inside the core, there happens a "point of change = 0". The T Field changes to S Field and moves outward once again. The T Field predominant entity we call earth lives and feeds itself in this way. The vigorous S Field sheaths, in full partnership with the material earth, creates outreach and environment for the wholeness of the living earth.

> **The outward breath of S Field is a flowing force field which contains all the vibrational information of the entity. The inward breath of the T Field is a flowing linear force field which contains and stores all the vibrational information of the entity. All energetic particles, planets, suns and cosmic bodies use this system of conscious interaction.**

The "points of change = 0" designate shifts from S to T Field or from T to S Field. They are Alpha and Omega. They are energetic places of essential cyclic change. The place and speed called zero is not an empty place but a dynamic event with conditions that have interconnections with adjoining field octaves.

The effect of the inward and outward breath is not that of a circuit. The way you can best think of the field activity is to compare it to the leaves and stems of a plant. Or if you prefer, you may think of the design as a flower with many petals, such as a rose, or a lotus, whose mirrored image reflects in the quiet waters of a pond (See **Illustration 3-1**).

A plant grows upward from the ground with the aid of the expanding S Field. It reaches its maximum growth. T Field follows the veins downward, grounding it to earth, providing the plant a way to utilize the energy of the sun, and providing many different paths to environmental adaptation. The tip of the branch or the leaf is a "point of change = 0". The plant has an Alpha and Omega. A plant is a good example of field activity because a plant uses fields directly as a source of life.

The unfolding of the petals of the lotus is an appropriate philosophical metaphor describing the full vibrational intelligence available within the scope of gravitational energies. The designs of MIND are continually dramatized by the forms of nature that live and die within the many environmental resources of your blue planet. The story of fields is unceasingly retold in new ways for your recognition and understanding of the laws of ONE. MIND is restless in its endeavors to reach artistic ideals and interesting recombinations of life on one planet so that all the cosmos may share in the music of creation. Nature always envisions the LAW as nothing can stand aside from the LAW. The answer to any question at all is YES. The LAW affirms ALL THAT IS. Nature demonstrates ALL THAT IS. YES.

GRAVITATION IS THREEFOLD

Another aspect of the gravitational field serves the need for orbits and stability. **Illustration 3-2** shows how the solar-galactic axis individualizes its forces. The axis is tapped to bring power into the earth's core. Cosmic energy feeds from the axis onto the equatorial plane of the earth like eddies in a water pond. Because the earth's core is iron and has a hexagonal vibrational pattern, the energy of the axis swirls into six divisions within the earth called cells. These cells develop sub-axes parallel to the primary axis. They express the same energy but in smaller octaves . It is these sub-axes that generate circuit loops through which your moon orbits. The swirling energy called cells feed molten material to the core where field separation takes place. These cells effect volcanism and tectonic plate movement. Because the sub-axes are not centralized you do not recognize one central pole. You have no way to test for the higher ST Field octaves or for USUT Fields except as they demonstrate CLOSURE and wave propagation.

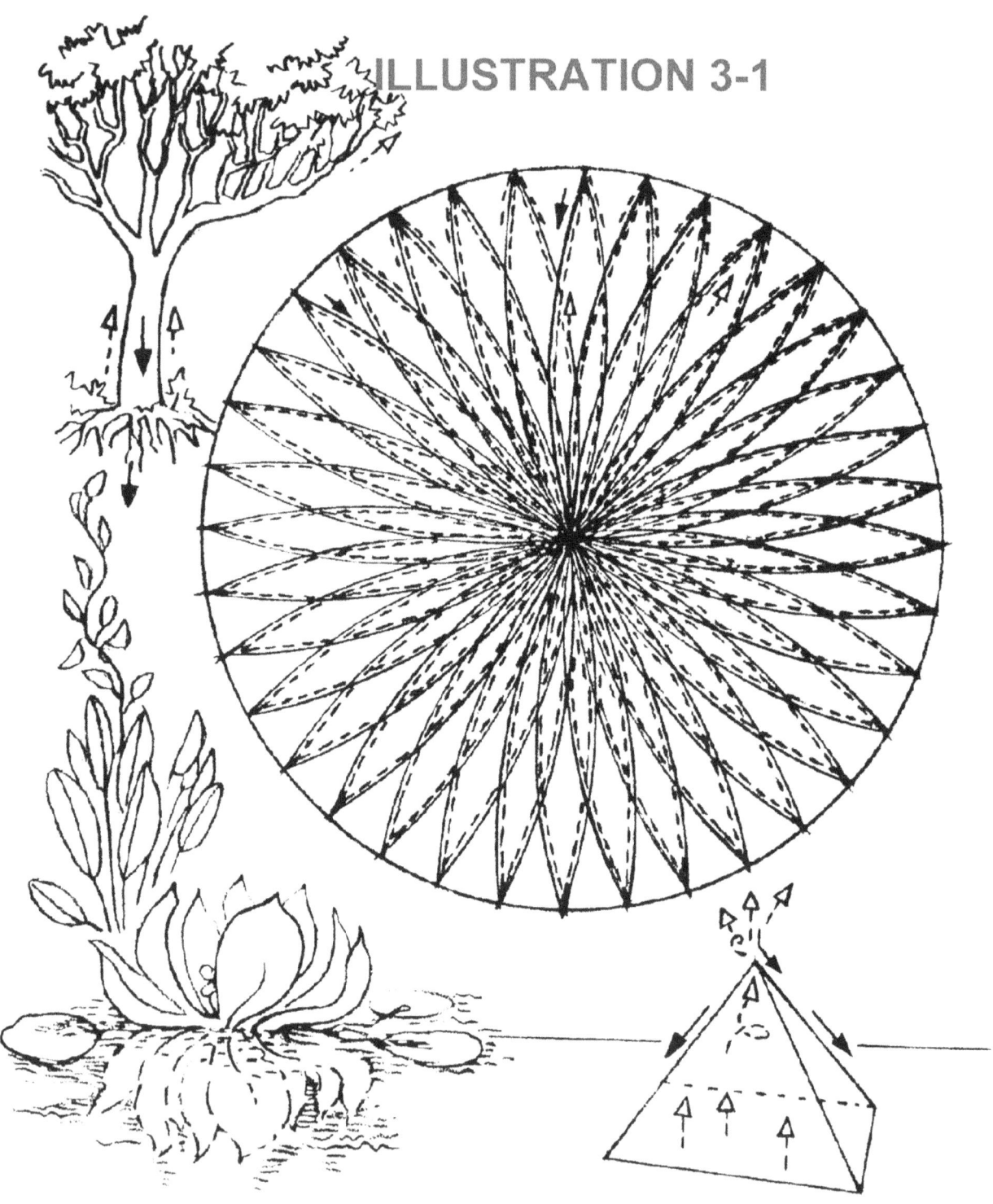

ILLUSTRATION 3-1

GRAVITY BREATHES IN AND OUT SIMULTANEOUSLY LIKE ANY LIVING ENTITY. FOR THE PEOPLE OF EARTH, GRAVITY SEEMS TO TRAVEL UP AND DOWN.

The earth's core stays in place because it is responsive to the solar-galactic axis. Yet it can shift and rotate. The core rotates faster than the earth's crust. The interdependence of the core and the crust is modulated by the energy within the cells. Sometimes the core flips with regard to the crust and causes minor havoc. The core is densely substantive. Its duration has requirements of elemental arrangements, similar, but different from iron, and different from the materials it accumulates to it.

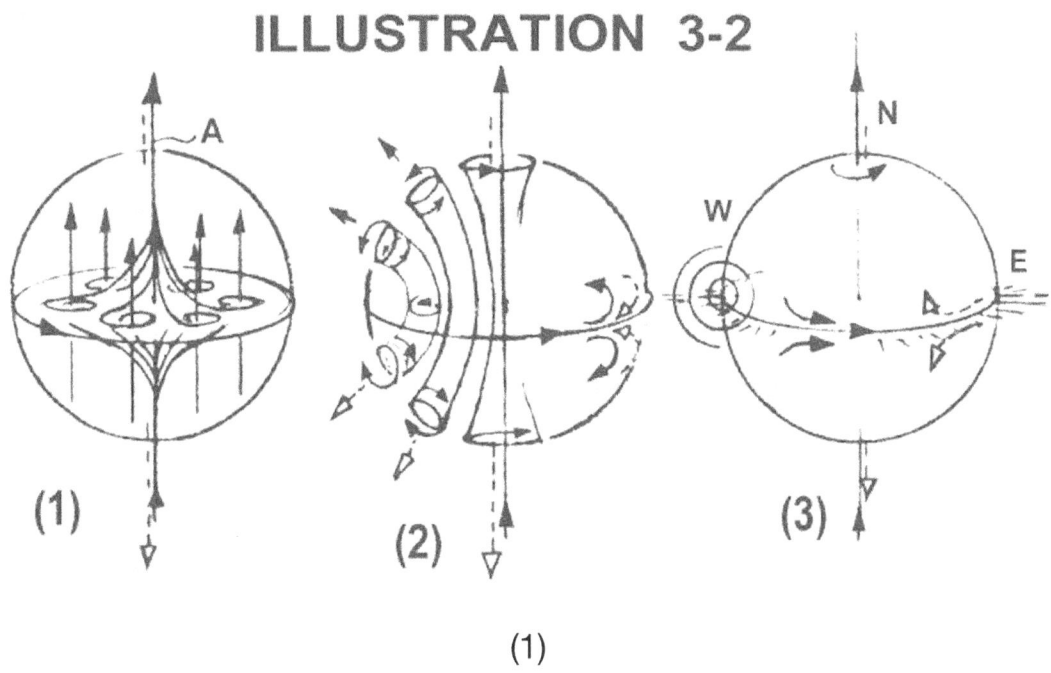

ILLUSTRATION 3-2

(1)
The solar axis (A) forms eddies upon the equatorial plane which establish dipolar circuits in the molten core of the earth.

(2)
These field circuits loop outward like any flux to establish the full diameter of the gravitational sphere. The circuit loops create, at their collective centers, many orbits including one in which the moon travels around the earth.

(3)
The earth revolves west to east, and so do the T Field currents around the equatorial plane (which lies 5 degrees tipped to the solar ecliptic). The S Fields flow the opposite paths, driving the oceans currents. S Field splays forth strongly at the equatorial plane (alternating with linear T Field gravitational lines). The earth has two equatorial planes, one established long ago relative to the rotational axis, the other, perpendicular to the present solar axis.

The core itself is responsible for the "magnetic" pole. Its iron-like material gives it properties you could call the "granddaddy of all magnets." This field is very different from the gravitational field and has a separate axial tilt.

We have described gravity in three different ways here; one, as breathing in and out, two, as feeding T Field from the mixed fields within the molten outer core, and three, as cellular eddies that form individual axes within the molten outer core. We can quickly compare two processes to an animals' need to breath and also to eat food. Sustenance has more than one requirement. The circuit loops formed by the sub-axes provide orbits for the moon and circulation for cosmic fields through the body of the earth.

The fuel for the planet's life comes primarily from the sun's circuit loops and solar winds. While planet earth has its own octave personality, it is always tied to its 'parents' apron strings. The sun provides both the orbit and the lighted energy that makes life possible on earth. Even the most primitive plants and animals feel the mother-father presence of the sun. As the sophisticated person sees farther into the sky and can know what he/she cannot see with eyes, the distant cosmic family can be better appreciated. And, we may add, the cosmic family will be eventually known as co-operative partners in the earthly celebration of life.

Whenever earth forces are considered, the influence of orbital and rotational speeds come into the picture. Formulas for "angular momentum" have been important in your calculations. To understand earth's speed in its orbit you will need to look at a simple model. **Illustration 3-3** will help you track orbital activity through circuit loops.

Circuit loops describe a torus. Energy travels the lines round and round in loops, but the torus formed from the loops stack themselves one upon the other with S Field separating each torus. (The electric wire windings of electric motors use the same procedure, each wire being coated with insulating paint.) Therefore, looped lines close to a cosmic source, such as a sun, are shorter. Their loops have a smaller diameter and travel less distance around a smaller circumference. The largest circuit loops are much longer and travel great distances around to complete the torus. Each torus has circulating energy centers known as orbits. Larger tori have orbits with greater diameter. Smaller tori have orbits with smaller diameters. Each orbit is insulated from the other. All of the circuit loops transit through the center of the sun's axial pole and **ALL ARE DRIVEN BY THE SAME SOLAR ENGINE AT THE SAME SPEED**. It is easy to estimate that energy, traveling at the same speed, will circulate a shorter distance in less time. Planets close to a sun make more revolutions around the sun (per period of time) than planets whose orbits are greater distances from the sun, and the slower will be their revolution around the sun.

A planet revolves on its own primary axis as a result of its early formation. Any axis is a section of a cosmic circuit or radial avenue. Any line of T Field flow will have a spiraling S Field sheath. The

ALCHEMICAL MANUAL for this MILLENNIUM

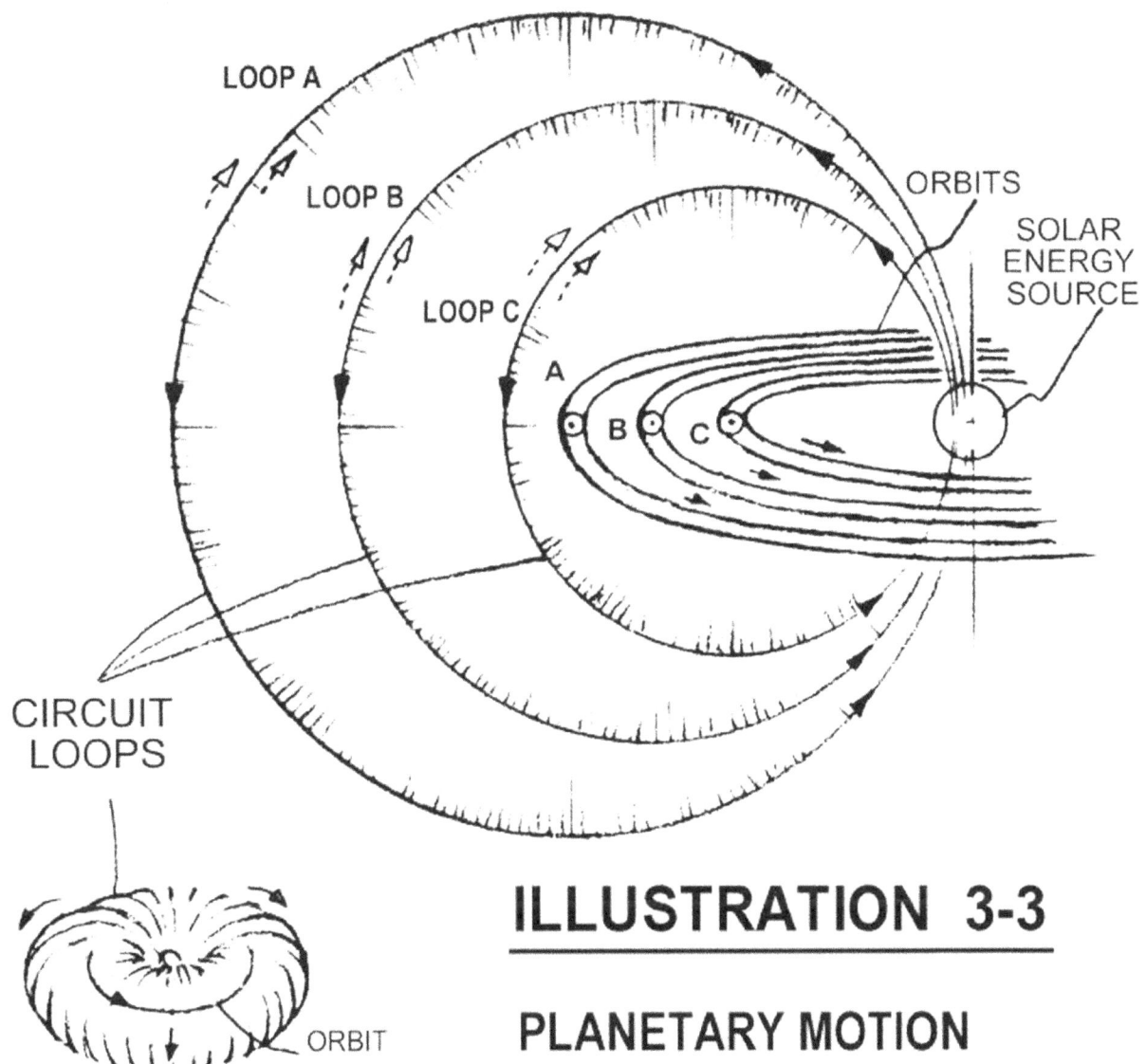

ILLUSTRATION 3-3

PLANETARY MOTION

ALL CIRCUIT LOOP CURRENTS ARE DRIVEN AT THE SAME SPEED BY THE SAME ENERGY SOURCE. THE SIZE OF THE LOOP, (ITS TRAVEL DISTANCE), DETERMINES THE RATE OF FLOW OF ITS ORBIT.

ORBIT C WILL COMPLETE ONE REVOLUTION BEFORE ORBIT B DOES, AND ORBIT B WILL COMPLETE A REVOLUTION BEFORE A.

INNER CIRCUIT LOOP EXPERIENCE MORE PRESSURE THAN THE OUTER ONES, (C MORE THAN A). THERE WILL BE POTENTIAL DIFFERENCES BETWEEN THE LOOPS. ALL THESE FACTORS DETERMINE THE RATE OF FLOW OF THE ORBITAL PATH.

ORBIT

CIRCUIT LOOPS FORM A TOROID AROUND AN ENERGY SOURCE

velocity of that spiral in conjunction with the potential within the flow (of the cosmic circuit) will show a characteristic "frequency" . You can think of any primary axis of having a "frequency" of travel. That spiraling force pushes material-in-formation into a spin. Once the spin is established, it is maintained by the axial spin and by "momentum of centrifugal force." Both "momentum" and "centrifugal force" will soon be discussed in detail.

We shall proceed to discuss many mechanical aspects of the gravitational field. Because gravity plays such an importance in daily life and is so often mentioned, we will abbreviate the Gravitational Space Field to GS Field, and abbreviate the Gravitational Time Field to GT Field, or together to GST Field.

GRAVITY PUSHES UPWARD AS A WIND AND COMES DOWNWARD AS A LINE

The in-out motion of the GST Field is continuous, as an eternal chorus singing an unending song. The upward spiral of the GS Field, is a windy flow. The GT Field begins its descent as a waterfall and reaches earth in multitudinous lines. The lines are very real. Each line surrounded by GS Field is a real octave force. The lines can be counted and the count varies from place to place. Each GT Field line of force travels on a gravitational radius. The force is a moving field, slowing as it approaches the earth's core. At any given altitude its speed can be measured. The slower its speed the more CLOSURE force it will exhibit.

The upward GS Field wind surrounds the GT Field. It ascends and accelerates. Its volume expands and expresses force in spiraled and radial patterns. GS Field force is tremendous and has been largely overlooked as an equal and opposite aspect of gravity.

Through the mechanics of ST Field separation the advantages of both space and time can be individually utilized. You have noticed how easily you can fall downward. You may want to also know that you can fall upward.

GT Field will pick anything with a T Field predominance and pull it to earth, to the ground, and on to center. Coils of expanding S Field gravity flow like a wind upward toward the outer circumference of the sphere, and push with it anything with an S Field predominance.

While gravity is a slow moving field it penetrates all materials of earth. You would expect that since it was gravity that brought all those materials together to form the earth. As the GT Field penetrates into and through materials it slows down even more. It becomes clear that both GT and GS Fields show preference to easy passage through materials. They will change position to avoid stumbling blocks. GS Field will choose to escape from a crack or a hole before it will tackle passage through basalt or lead. GT Field lines will take a host of circuitous routes before it will jump through air or helium, rubber or plastic. GT Field lines do not prefer to be deviated from their usual course by more than 45°, yet they can be coaxed to make easy turns.

It takes *time* for GT Field lines to penetrate dense materials. Sudden movements perpendicular to the flow of the GT Field can interrupt the flow. Objects or gases of high speed can interrupt the flow. Explosions of S Field can lift rockets off the ground and lift them to a place of gravitational escape. Tornado winds can thrust GT Fields aside. The release of compressed S Field from an atomic bomb blows GT Fields asunder . Yet be assured that any presence of a quiet but appropriately strong S Field will part the lines of gravity like a curtain on a theater's stage.

S Field and T Field can be separated, each acting independently, but for a very brief time. To sustain an independent field takes constant management of forces. One explosion must quickly follow another. However, durable field predominances can be established in materials. The study of field separation will occupy your scientists as soon as they realize its mechanical and chemical advantages. Our engineers use separated S Fields to place a protective envelope around our spacecraft to shunt off gravitational lines. Our advantage is that we have very dense elements from which concentrated S Fields can be liberated. Overtoned harmonies must be matched within materials to emulate the harmonies of earth's gravity in order to have any effect on gravity. As our discussion proceeds concerning basic mechanics you will be given a greater grasp of this subject.

A LIVING FIELD BEGETS LIFE

The earth, its gravitational field, and "Mother Nature" are all ONE, all parts, all wholeness. To understand fields you can look to plant life as your first witness.

The most important prototype for earthly field separation and field integration is a tree. Into a living field, an idea is projected. The idea is detailed, and clothed, coded and synthesized, into a seed. (There is no question that the egg came before the chicken). The seed of a tree was developed by a living being only after the IDEA of a tree was completed in full cycle. Ideas are powerful but not substantial (as are ST Fields) It is ideas that signal the fields into precise construction. That can occur only when the nature of fields is fully understood.

A seed, in a place of nurturing , grows up tall with a single trunk, then divides to fill available space. The assertive "life" of the tree climbs up the GS Field like a ladder. The center of the trunk walks the GT Field downward to be grounded . As long as GT Field follows that downward path, the GS Field will spiral upward around it. The tree's cambium layer grows up and out. Water at the roots enhance the fields grounding. Water is drawn up into the cells of life to provide nourishment in the forms of oxygen and minerals to the whole tree. Each branching of the limbs gathers in more GT Field to be carried downward. Each leaf mimics the pattern of the expansion of the GS

Field. Each vein in the leaf draws in more GT Field to the trunk. The spreading leaves and branches gather electrons and photons from the sunlight and solar winds as nourishment. The tree balances in the directional gravitational fields to fully utilize the field flows.

Trees gather GT Field lines into their trunks. The "heavier" the tree the more lines it will gather. The tree is GT Field predominant. Between the trees in the forest there are fewer GT Field predominant lines. In that space animals found homes and thrived. **The design of all animals, humans and angels favors a GS Field predominant environment.**

Acids show a GT Field predominance. Alkali show a GS Field predominance. Every life has a preference of formulated predominance. Each life form is integrated into an appropriate environment, and cannot be separated too far from it. The idea of life can be sent far and wide but the form of life cannot. The form is integrated to a nurturing place.

No man, nor animals, live well in direct sun or ultraviolet rays. The earth itself has enjoyed that its forests cover the land, just as a man enjoys the hair on his head. The forests convert carbon dioxide to oxygen as they breathe. Without forests, man and animals would have no good air to breathe. The evaporation of bodies of water puts droplets of water vapor into the air, but not oxygen, the requirement of breath. Without trees covering large portions of earth, life as we know it could not exist. All chemistry develops from the interplay of fields. Intelligent beings are obligated to understand that chemistry, before they commit acts to destroy the blue planet by deforestation.

Configuration of field predominances are part of your daily life and your history. A peaked roof on your house not only sheds water but shunts GT Field lines to your walls and allows your dwelling to gain GS Field predominance . Rubber wheels on your vehicle are GS Field insulators and can prevent your being struck by lightening. Dampness attracts GT Field lines and is bad for your health. Reeds filled with air pockets and woven into a raft will float you down a river, adding to your GS Field buoyancy. You feel better in barometric "high" weather than in "low" weather. The expansion of S Field gasses from yeast rises your bread.

Experiments in pyramidal structures show a benefit to growth in plants and animals. Inside a pyramid, the GS Field coming up from the ground, is circulated in frequencies that favor good health. The space discourages destructive bacteria from growing. Add sunlight to a pyramid and you will grow bigger plants. Those people who use a pyramid for meditation say that the space enhances their mental abilities to alter frequencies. The ancient pyramid builders used the principles of the GST Fields to make a tomb for the pharaohs that would guarantee preserving the body and maintaining the soul. The massive pyramids were the attempts of kings to separate fields by creating cave-like places within heavy materials. The rock would divert the GT Field lines away from small spaces of strong GS Field predominance where a dead king would be entombed, where decay would not occur. The plan worked successfully for the kings but not so well for the workers. The kingly descendants of Egypt had a traditional knowledge of fields understood only by the avatars attached to the throne. That remnant of knowledge quickly died out as Egypt was overrun. It is our opinion that a truly advanced civilization would not have constructed pyramids of this design. Field separations for the deceased person requires that the flesh be discarded with all its cellular DNA, and that the soul be released to freedom. Resurrection of the flesh is possible but not the best way to go. Buddhist and Hindu practices of separating the fields by burning the bodies of the dead without ritual paraphernalia is more effective for spirit liberation.

Transcendental meditation practices demonstrate that the body can be surrounded by swirling GS Field through sustained concentration. The result being a levitation of the whole seated body into the air briefly. In this process GT Field lines are parted to the sides of the bodily sphere.

Eons ago the ape discovered that by standing upright he encountered fewer downpouring GT Field lines. It took less effort to walk upright than ambling on "all fours". Caves may have helped the spiritual development of early man, as caves often show a strong GS Field predominance. Pockets and cleavages in rock are GS Field predominant. Hermits seeking enlightenment often took to caves to look for GOD within themselves.

On the other hand, wisdom seekers climb to mountain tops to attract stronger GT Field lines which transit their spine, thus wrapping their organs in stronger GS Field spheres. Sending and receiving messages is effectively accomplished from mountain tops. A sphere can enlarge without interference in an open space.

Water has been called the birthplace of life as you know it. Water is a middle octave, a balance of fields that allows life to be mobile rather than being attached to the ground . Water allows for greater decision making and greater influence. It was the first environment for life's development. Primitive animal life began as bubbles in a quiet sea. An alliance was made fo r consciousness with ALL THAT IS.

Flowers and seed heads, the fruits of a plant are designed to match the reproductive capacity of the fruitful gravitational fields of the planet. An apple or an orange looks like the GST Field looks. Their rounded forms, protective skin, segmented meridian divisions, their nourishing pulp, seed cases, tiny seeds and axes all describe the GST Fields. Earth establishes multinumerous varieties of seeds intended to distribute themselves throughout the Universe. The seeds of earth will be carried far and wide. The best seeds will flourish in new soil with new opportunities for development. Earth is a garden of beautiful life with a WILL to continued growth.

Within the breath of gravity is the sound of its heart beat. The body of earth pulses with energies from the sun and galaxy. If you listen, you will hear it in your heart as well. Microwave sensitive telescopes can record the beat. The solar engines turn ceaselessly to pump the currents of its GST Field through the giant circuit loops that, in turn, carry the planets steadfastly in their orbits.

Can you imagine that a rocky, watery planet could have a soul, and that such a soul could be released by a spent body? Can you imagine that the youthful soul of the planet Mars might be dwelling elsewhere, reestablishing its essence on more fertile soil? Souls, like seeds, blow on the winds. The chemistry of spirit is not as restricted as flesh and blood.

Think of a sci-fi scenario where earth was shortly to be struck by a giant comet. Comprehending a collision course, the people of earth could calculate their extinction, and then make a radio appeal to all Universal stations to accept the souls of earth. Each galactic station could put the case before their government asking, "Do we want those beings here as refugees ? Are they worthy to be among us? Shall we send a travel beam to receive them or shall we let them be cast adrift?" Would it be an easy decision, or would the soul seeds of earth be turned away? Who would be taken? Who would not? Would it be a collective agreement? What would be the terms? This metaphor should make you think about who your are in a cosmic sense, and what your relationship to the garden of earth might be.

In Santa Cruz, California, where deep beneath the surface of the earth are a network of fractures from earthquakes, there is a very special place in the countryside where the farmers stand crooked. The owners of the land, where compasses go crazy, set up a commercial showcase for the gravitational field anomaly. They call it "The Mystery Spot". Deep within the earth GS Field collects and drives upward, swirling around extremely strong GT Field lines that travel at odd angles. The owners have rigged a collection of demonstrations to delight the dizzy visitor. One demonstration that is true and clear shows that in one spot a standing person looks bigger than in a nearby spot where a person appears smaller. The photographs in our **Illustration 3-4** are not doctored. You may visit and see for yourself.

> **We want to point out that a strong GT Field will shrink the corpuscles through which light avels while a strong GS Field will enlarge the corpuscles through which light travels.**

In the photograph the man who appears larger is standing in an extremely strong GS Field, the man who appears smaller is standing in an extremely concentrated GT Field just adjacent. Notice the carpenter's level and balanced golf ball on the ground.

The phenomenon of the expanding corpuscles of the GS Field has been utilized in the designs of electromagnetic microscopes. The basic designs employ a very strong current in a dipole ring whose lines of force cause the S Field in the center to splay. Put in series, these rings continue to enlarge the smallest lighted detail of the microscopic world.

Illustration 3-4

AT THE "MYSTERY SPOT" NEAR SANTA CRUZ, CA. PEOPLE STANDING AT DIFFERENT PLACES APPEAR TO CHANGE SIZE. STRONGLY DIFFERENTIATED SPACE-TIME FIELDS IN THE EARTHQUAKE AREA CHANGE HOW LIGHT IS TRANSISTED TO A VIEWER OR A CAMERA.

THE EXPERIENCE OF SIGNALS IN A FIELD ARE IN DIRECT RELATIONSHIP TO THE DENSITY OF THAT FIELD. IN A GRAVITATIONAL FIELD LIGHT, SPACE AND TIME CAN DIFFER IN A REAL WAY, EVEN IN A SMALL SPACE, DEPENDING UPON THE GEOLOGY OF THE AREA.

NOTICE HOW THE CARPENTER'S LEVEL AND GOLF BALL ARE PLACED TO DEFINE A LEVEL BOUNDARY.

A TRIP TO THE "MYSTERY SPOT" CAN BE REVEALING. NATURE'S PHENOMENA CAN BE FUN, AND YET, NEVER TO BE TAKEN FOR GRANTED.

At different episodes of the life of a planet, the gravitational output shifts in predominance. During a period of time when the GST Fields express a GS Field predominance the growth of cellular life on the planet shows enlargement. The result is bigger plants, bigger animals, bigger insects, and bigger bodies of water. Not only do these life forms appear bigger, they grow bigger. At that time there is less problem with weight in terms of exertion. Any sudden shift in the predominance and strength of a GST Field can cause selective extinction in biology.

You have been living in a strong GT Field predominance now for about 50,000 years. A change in predominance is due for earth. As a planet grows older and hardens its core, it cannot return to a truly buoyant youthful stage. As a solar system moves through a galactic orbit it is subject to alternating predominances that characterize any orbital ring.

Deep meditation practices develop the abilities of some students to levitate objects or to levitate themselves. This is done according to laws of physics and is not a special Godly gift. It is done by focusing GS Fields through the body and placing high tori concentrations around an object. This causes a parting of the GT Field lines and allows the object to fall upwards. When the concentration is disturbed, the GT Field lines again attach to the object and it falls.

That exercise is valuable because it shows you that your body and brain have abilities to manipulate fields in ways you have not believed possible. Belief and desire initiate adaptation.

Practices of using CHI in medicine and in the martial arts of the Orient have demonstrated innumerable times that humans have a wide range of ability to alter conditions of life and motion. It is the purpose of this treatise to change your belief system so that you may experience more of your own capabilities.

Maneuvering CHI within the discipline of the medical arts has had brilliant success. The work needs expansion and scientific understanding. Work in this field is worthy of your study. Its teachers present themselves all over the world. Traditionalists within all the scientific disciplines need to interchange information and data, as well as respect, so that TRUTH can emerge.

THE EXPERIENCE OF GRAVITY

You are cognizant of the presence of gravity every moment of every day. You are aware of its downward pressures, with falling bodies and with weight. Gravity shapes the earth, all that is in it and on it. You have chosen to ignore the fact that gravity also pushes upward, and that gasses are constantly rising. Therefore your formulations of the mechanics of GST Fields has not included the equal and opposite effects of rising space.

It is nearly impossible to speak about gravity without considering the abstract view of force itself. One force is forever tangled up in another. Chapter 3 and Chapter 4 must be cross-references to be understood. We shall begin by creating a visualization of gravity as it acts within your world.

FALLING BODIES

A falling body is said to accelerate in a free fall. It is said that all bodies fall at the same rate. This explanation begins by knowing the constant speed (more or less) of the GT Field lines. The lines coalesce close to the earth's crust, about 20 miles up. As they delineate, they become directional, following radial paths. Their vibrational speeds are appropriate to pull with force any materials capable of CLOSURE with earth's core. These GT Field flows, as the lines descend, are decelerating in their approach to center.

If you were flying in an airplane 10 miles up, flying perpendicularly to the GST Fields, and you dropped an object from your plane, at the place of release the object would be traveling at zero rate with regard to the speed of travel of the river of GT Field lines. The downward force of CLOSURE would try to penetrate the object and would meet resistance at the object's surface tension. The surface tension acts as a wavefront distributing the force over the surface and allowing the GT Field thrust to move the whole object and not just a part of it. The surface tension protects the integrity of the object.

How can an object resist such a force? By non-resistance. It must become one with the force. Inside the object the GT Field lines begin to CLOSE with each T Field point in the material. With each speed gain the internal ST Field of the object must seek a new balance. Kinetic S Field must be equal to potential T Field. Potential T Field must be equal to kinetic S Field. Together the S and T Field must have a speed ratio like the GST Field in which it finds itself. As GST Field penetrates each field point within the object; the object gains speed and stores up "mass".

As an object gains speed, the differential in speeds between the GST Field force and the object's ST Fields force lessens. For instance, a force traveling a constant 100 mph has to work harder to push an object traveling in the same direction at 5 mph than to push an object traveling 50 mph. When the object travels at 50 mph the constant force is meeting less resistance. Therefore, the same force can push the object faster in terms of mph. Thus a falling object accelerates as it gives up resistance to the GST Field force flow in an incremental manner. When the object reaches the same speed as the GT Fields linear flow the object no longer accelerates. You seldom see this as it takes a long time for the object to reach the speed of the flow. (Sky divers do experience the place where acceleration ceases.) The falling object strikes the ground in shocking impact. As it gives up its speed and newly acquired "mass", a new field balance is achieved for a new speed ratio.

It is the same idea as if you put a wooden toy boat, carrying a toy soldier, into a rapidly flowing river. The boat would accelerate in the water until it reached the same rate as the water's flow. The safety of the boat and its soldier requires that it not resist the overwhelming force of the water's flow, except at its surface tension. Since it can not beat the force, it joins it, adjusting to it incrementally. You could say that the ST Field of the toy boat and soldier joins in CLOSURE with the ratio of the GT Field/GS Field that surrounds it.

If you were talking instead, about a man falling in an elevator in a gravitational field you would have the same phenomena present. The man and the elevator within one unifying sphere and surface tension make internal adjustments to any speed change. A living being can make these changes within certain limits of time and space before the integrity of the body is given up. The adjustment of the S and T Field ratio takes place at each juncture of the GT Field line and T Field point and at the center of gravity of each unifying sphere.

The fact that a traveling object gains "mass" is not relative to the acceleration of its fall in a GST Field. It was pointed out that objects of variable "mass" fall at the same rate. The reasonable discrepancy here is in your definition of "mass", as we will soon show.

> **The speed at which any object travels measures its kinetic energy. That exact speed (energy) must be balanced by the potentiality of its T Field points inside the object. As speed changes so does its internal potential (energy). S Field >< T Field.**

This LAW was expressed in Albert Einstein's famous formula $E = mc^2$. The balance applies to any speed and any pressure. A traveling body gains "mass". Any object is surrounded by T Field corpuscles and has no trouble using its resources. That is to say, the extra "mass" is drawn in from the surrounding fields.

An object falling in CLOSURE in a gravitational field gains "mass" at a uniform rate. Any T Field point goes through a spiraled process of energy storage. It takes its *time*. T Field points, traveling at the same speed, are alike. They are the basic particle characteristic of the octave. Being the same, all T Field points absorb energy at the same rate. The number of points in an object has little effect upon the rate of absorption. Therefore, under the same octave conditions, all objects CLOSE (fall) at the same rate of acceleration. (The *weight* of an object depend wholly upon the number of T Field points it has in its body).

Any object surrounding itself with a surface tension acts as a single, if irregular, sphere. The center of that sphere carries a juncture of T Field points (mass). As you are familiar with a point called the "center of gravity", you are aware of a centering sphere. That point of accumulated T Field could exist in the middle of an empty box. It need not settle in T Field points within structures (such as atoms).

The ST Fields always unify essences by overtoning in encompassing spheres. These spheres are really present and always act to support forces in motion. A gravitational field recognizes any unifying sphere in its octave and responds mechanically to it.

> **The force of the GT Field is equal to its speed and the number of lines acting upon all the points within a T Field predominant object in its octave.**

This accounts for the wide variations in the gravitational forces of cosmic bodies. The equal and opposite action of a rising body in the GS Field follows the same pattern of acceleration. The body falling upward is generally in the form of gasses. These gasses may find a level of buoyancy to drift at high altitudes or they may escape the gravitational field into outer space. Or if lines of GT Field are parted any object will fall upward in the GS Field. It is reasonable to suppose that most S Field predominant objects have already escaped into space unless they are trapped within a structure. Most S Field predominant life forms are invisible. That being the case you are not very aware of the balance of objects in earth.

A fast traveling object moving laterally to GT Field lines can avoid the penetration of those lines. CLOSURE is a slow process. It takes time. T Field points must first identify the GT Field octaves, their vibratory rates, and their speeds. Secondly, they must make an agreement to integrate into one point. Both directionality and surface tension can discourage CLOSURE. This has given engineers a positive advantage in designing aircraft.

An aircraft uses thrusts against air to gain speed. It is that speed that keeps the craft from falling in the GT Field. A hovering craft, such as a helicopter, not only thrusts against air, it disperses field lines, but it is less efficient than a moving craft.

The knowledge of the mechanics of fields is part of your genetic heritage. You intrinsically know that if you want to jump over a big hole, you get a running start. Lateral speed will keep you from falling. You also know that T Field points store more GT Field at a fairly slow rate, and also release that storage at a fairly slow rate. When you get up speed you can hold that speed just a little while. Both a little speed and a little time will get you over that big hole. Every child learns it for himself.

WHAT IS WEIGHT?

When an object rests on a spring scale, you notice that the needle gauge constantly reads a specific measure for the same object. You say that object has a characteristic weight. Without gravity the object is weightless and floats about in space. How is the flowing force of gravity acting upon the object? Weight is a mystery that you are involved with every day of your life. Let us show you how to unravel this mystery.

You pick up an axe in one hand and a stick of about the same size in the other hand. You say that the axe is heavier that the stick. Gravity pulls down on it harder than it does the stick. Yet you also know that these two objects fall at the same rate. What makes materials act differently in gravity? Every material is made up of organized and bonded arrangements of mixed fields. Each item has a specific count of T Field centers and S Field spheres vibrating at special speeds under pressure. It is ST Fields that comprise atoms, molecules, and the spaces in between these units. The material is packed together in surface tension. There are far more T Field points packed together within the axe than within the stick. The number could be counted exactly, even though the numbers are in the billions.

The GT Field has more lines per square inch than the T Field points of any material. The resting condition of any object allows time for the gravitational field to penetrate it on its way to the center of the earth. Any single T Field point within an object will engage at least one GT Field line.

A T field point picks up a GT Field line and swirls it into its center. Having taken from it an appropriate mass, it releases the line again so that it continues its path to earth's center. It swirls inward and swirls outward. Once released, the GT Field line is traveling slower than before. GT Field lines that are not engaged by T Field points continue their fast travel through the object to ground. The released and slower GT Field lines now behave in a noticeably different manner. See **Illustration 3-5**.

> **As GT Field slows down it shows an increasing ability to CLOSE. The lines then have a similarity to electricity because they choose to follow a circuit. There will be an exact number of slow GT Field lines to match each T Field point in the object. The force of attraction to the earth's center is an additive collection of slow GT Field lines, one for one, with the T field points of an object. This happens consistently. This shows you clearly that a T Field point is your basic particle and constitutes the primary building block of material accretion for your octave.**

The stick has fewer T Field points than the axe, fewer GT Field lines are encountered and slowed. There is less accumulation of lines that pull with the earth, and therefore, you say the stick weighs less. In addition, the stick has an exceedingly large number of S Field predominant spheres within and around its cells. S Field ignores the GT Field lines and tries to fall upward. This adds a counter force to the pull of the slowed GT Field lines. The stick is said to be buoyant, and will float on top of water. Buoyancy will be discussed shortly as it is the reciprocal correlate of weight.

When the GT lines are slowed down by the T Field points in an object they take on a new important feature. They can be contained by an S Field insulator. They seek out and follow a circuit. That circuit may take the GT Field lines up and down and all around in order to form a circuit 'ground' to the earth. They follow a "conductor" material.

Imagine a friendly farmer carrying a bucket of water. The farmer knows that he usually weights 200 lbs. But as he steps on the scale with his bucket of water in hand he weights 230 lbs. The extra weight of the bucket and its water is recorded on the scale because the GT Field lines, now traveling slower, have traveled up the handle, up the farmers arm, through his muscles and bones, and out his feet to the scale. They are then carried through the scale's gauge to the ground .

ALCHEMICAL MANUAL for this MILLENNIUM

ILLUSTRATION 3-5

GRAVITY IS AN ST FIELD CIRCUIT THAT FLOWS LIKE A RIVER. ITS MULTIPLE HARMONIES ALLOW IT TO PENETRATE AND ENGAGE WITH ANY MATERIAL ON EARTH. THAT TAKES TIME.

THE SPIRALED ENGAGEMENT DANCE SLOWS THE SPEED OF THE T FIELD LINE, MAKING IT MORE PARTICULATE, INCREASING ITS POTENTIAL TO 'CLOSE'.

GT FIELD LINE NOT ENGAGED

GT FIELD LINE APPROACHING CENTER OF EARTH

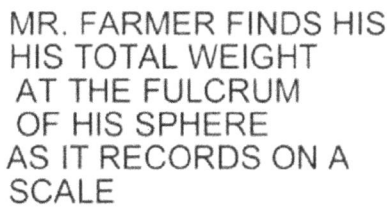

GT FIELD LINE ENTERS A T FIELD POINT (CORPUSCLE) WITHIN MATERIAL

GT FIELD LINE EMERGES FROM POINT TRAVELING SLOWER. THEN IT ACTS LIKE AN ELECTRICAL CIRCUIT

SYMBOL FOR 'GROUND',

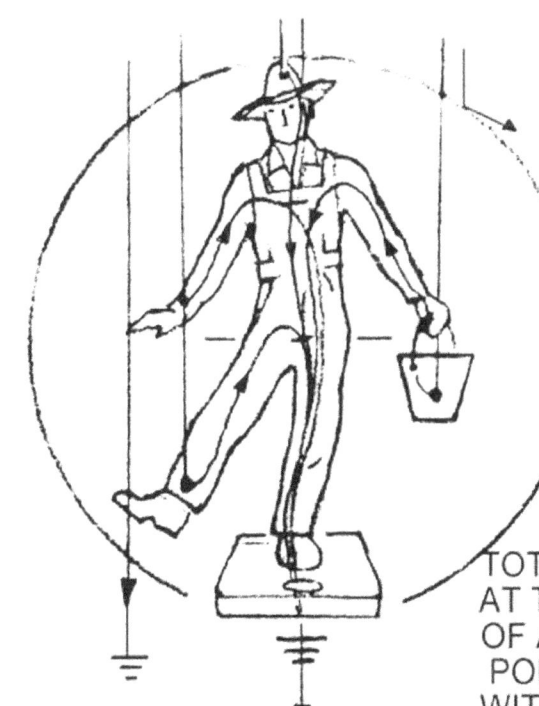

MR. FARMER FINDS HIS HIS TOTAL WEIGHT AT THE FULCRUM OF HIS SPHERE AS IT RECORDS ON A SCALE

THE SLOWED GT FIELD LINES USE THE BODY AS AN ELECTRICAL CONDUCTOR, SEEKING THE EASIEST PATH TO 'GROUND'

TOTAL WEIGHT IS MEASURED AT THE FULCRUM OF A COUNT OF ALL GRAVITY OCCUPIED T FIELD POINTS (CORPUSCLES) WITHIN A MATERIAL

Now the farmer does a trick for us. He stands on one foot on the scale. The scale's gauge reads the same. The weight of one leg travels upward to his spine and travels down the other leg, bringing the total weight (the slowed GT Field lines) through his one ankle to the scale. The farmer is the circuit and the conductor. He has had to adjust the position of his arms, legs, and spine to become a fulcrum for the center of gravity of his sphere.

That centering fulcrum is the center of a sphere of force surrounding and containing all parts of the friendly farmer and his bucket of water. The sphere goes way out beyond his hat and as far down into the ground. The sphere has divided itself to equalize the strength of force on all sides relative to the farmer's ankle fulcrum. He dismisses his miraculous trick by saying that any farmer can stand on one foot while holding a bucket of water. It is one of those common everyday miracles a person takes for granted. The farmer has not known that he creates new spheres every second of everyday.

Taking your spring scale to the marketplace you prepare to buy some potatoes. You can weigh one potato at a time then add all the weights together to know the price. Or, you can put all your potatoes on the scale at the same time and read out the total weight. The scale joins all the potatoes together in a common circuit, a unity of slowed GT Field lines, which show collectively on the gauge. You are counting and adding forces, one GT Field line to another. A new sphere surrounds the unified circuit. **ADDING is a hallmark of forces of the T Field and GT Field.**

When T Field points spirally interact with GT Field lines in free fall, the points are not connected in circuitry. Each line interacts with each T Field point at a specific rate. The effect of the force is not noticed collectively until the object strikes the ground. The unifying sphere plays a minor role in an object's free fall but a major role when the object strikes the ground.

Now, drop one of your potatoes onto your scale. Notice that upon strike your scale registers a much heavier weight than when you had rested the potato on the scale. But just after impact the scale rebounds to normal weight. You say that a falling potato gains weight, or gains "mass." A potato at rest keeps a normal, predictable weight. The greater distance the potato falls the more impact "weight" is registered on the scale. Something else, some other force besides "weight" is being measured with a falling body. It can be called "thrust," which means a new force being added to an old force. We have to refer here to the pages on falling or traveling bodies. Thrust is an aspect of S Field and GS Field. It plays a dynamic role in your experience with gravity.

LEVERAGE

A water pump with a long handle is easier to use than one with a shorter handle. Leverage plays a big part in " mechanical advantage". It has unending design applications in mechanical engineering. So we will give it a bit of attention here.

Gravity (GT Field lines) penetrates all the T Field points on the handle of the pump. The resultant GT Field lines being slowed down by the encounter of the handle have to travel across the handle to the pump spout to find a place to 'ground'. The new field quanta are slow enough to attract and gather new GT Field lines to them. Grabbing these lines, and again releasing them, they accrete the presence of "mass" into the handle. The longer the handle and the greater distance to a grounded fulcrum the more "mass" and weight the handle has. It has attracted exponentially more GT Field lines.

When the pump handle is used as a lever, like a radius of a large sphere, its outer tip describes a greater distance. The man pumping the water with a long handle has the advantage of distance over force and, using the heavier handle is assisted in his downward stroke by the increased gravitational pull.

Further detailed work on simple mechanics will be presented in especially prepared new works at a later time. In this survey we must set some detailing aside.

FRICTION

Friction occurs when T Field lines are broken and try to find new locations to CLOSE. When a big snowstorm breaks down the electric wires that serve your houses you can see power sparking off the ends of those wires looking for a new place to ground. You seldom think of that broken circuit in terms of friction, but it is the same phenomenon.

You know when you put pressure and speed between materials you produce heat and, sometimes, sparks. Much of that has to do with GT Field lines that join one material to the other in circuitry. CLOSURE and GT field lines act as linear bonds that join T Field points within objects in association, especially objects that are actually linked by the slow GT Fields in a circuit to 'ground'.

When we talk about circuits we indicate that there is the balancing condition of an S Field sheath around the circuits. That S Field is likely to be under some condition of pressure.

When you have a metal plate resting on a table you have the normal pressure of CLOSURE caused by the GT Field between the metal plate and the table. If you lean upon the metal plate you are increasing the force of CLOSURE. The T Field points in each material will show a potential gain. The S Field spheres around those points respond with equal and opposite kinetic force. GT Field lines that bond show increased potential flow.

Take a look at **Illustration 3-6**. The diagram will show the way that the GT Field circuits are broken each time one material moves over another. It shows that the union of GT Field lines and T Field points in the materials are disengaged with sudden movement between corpuscles, and then how they reengage with neighboring corpuscles, only to be broken again. Each time a GT Field line rips loose from a T Field center potential, it is directly exposed to an S Field corpuscle. The T Field line is suddenly confronted with an S Field barrier. It faces an S Field predominant gap (a sphere) that it must somehow cross. It backs up and gains quanta in order to make a leap. The T Field energy spins around and through a series of wavefronts. The S Field sphere is in compression and spinning with kinetic power. The spinning of a jumping spark can be compared to couples dancing round and round to a polka band of musicians. Couples join hands and spin around a fulcrum (a point of balance) as they move across the floor.

The confrontation of spinning fields emits (radiates) a range of characteristic wavelengths into the surrounding fields. Among those signals are wavelengths that you recognize as heat and light.

During a storm, air heavily laden with moisture and clouds becomes the place where T Field points collect. Each atmospheric T Field point latches onto a GT Field line (like any material object)

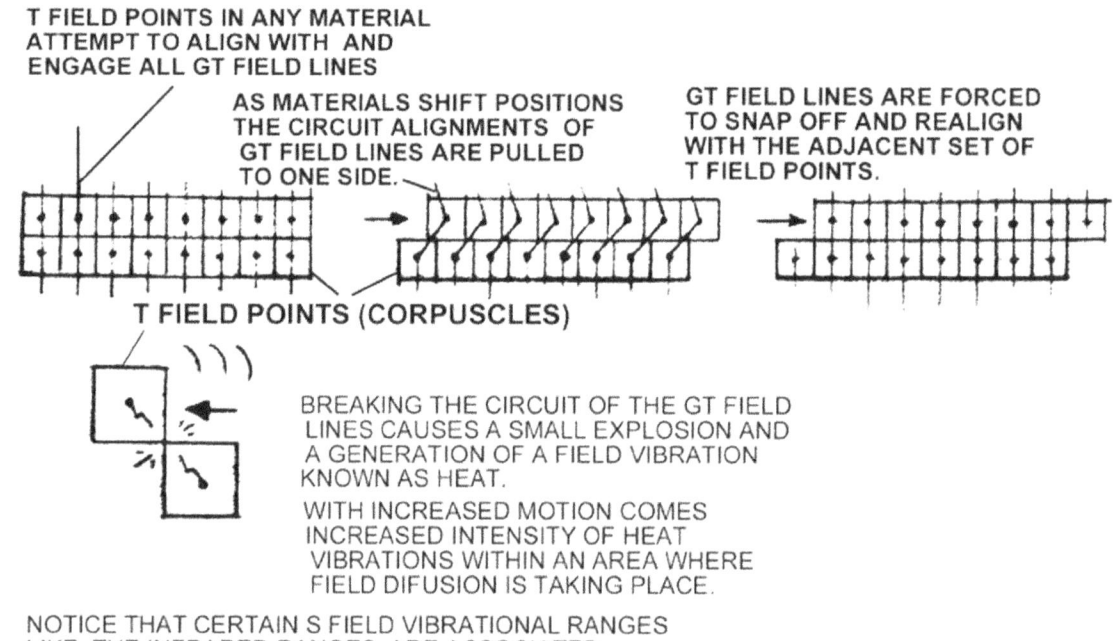

and slows down the line. This new slow GT Field line is of about the same quanta you call electricity. You have electrical lines running around clouds looking for a way to 'ground'. Between the clouds and the earth there may be an S Field predominant gap or perhaps falling rain (an excellent conductor). When a T Field concentration in the sky is gathering together to make a leap to earth, it informs the earth first by way of USUT Field signals. The shortest appropriate route to earth is agreed upon to facilitate CLOSURE. The earth attempts to provide a route for the GT Field lines to make their leap. The earth reaches out with a plume of GS Field to act as a balancing sheath. The outreach is equal and opposite to the voltage of the lightning strike. As the GT Field drives deep into the earth, the GS Field plume leaps high into the air above the clouds. Above the clouds it intercepts ions in the atmosphere and is often seen to glow with a pinkish light that has been called a 'sprite'. The sheath around the lightning bolt is radiative and you see a bright flash of light. A lightning bolt causes a great field disturbance by its surrounding waves in compression. This field shock translates into sound, the thunder that you hear after a lightning strike. In a storm, this phenomena of weight, friction and falling bodies all interact in a dynamic display of forces.

Painting by Ludolf Bachuysen 1667 **National Gallery of Art, USA**

BUOYANCY

The GS Field expands upward from the center of the earth. It pushes up through materials as well as cracks and fissures. GS Fields move up through S Field trapped by T Field bonds in the elemental structures of earth. It activates that S Field and inspires it to rise up. All S Fields thrust upward and out to complete gravitational cycle. Each material responds to this GS Field thrust in its own individual way. That material response is called buoyancy.

Buoyancy is the opposite equivalent to weight. What you know about weight applies like a reciprocal to GS Field expression in materials. T Fields always express linearly with strong directionality. S Field expresses spherically and multidirectionally. The measurement of buoyancy cannot be made with a spring scale, but with a standardized balloon, or flotation in a beaker. From your experiments has come a table of flotation measurements called "specific gravity," which measures material flotation relative to water. A comparative measurement of flotation will show you how much kinetic activity is trapped, or caught up in, a material. It will show what quantitative relationship it has with the number of T Field points in that material.

> **Any field, within or without an object (material), will seek and find the companionship of another environmental field of the same vibrational speed.**

A place of buoyancy, the specific environment in which a material floats, must be defined by a mutual compatibility of field speeds.

We must again point out that the speed of the gravitational field is slower near the center of the earth than at the furthermost circumference of the gravitational sphere. At any point within the GST Field the speed of the downward force is equal to and opposite to the upward force and speed. You can say that there is more pressure at the center of the earth than at the circumference of the gravitational field. At sea level the GST Field is more T Field predominant. In the upper atmosphere is increasingly S Field predominant.

The gravitational field is divided into concentric rings according to the prevalent harmonic octave structure, based on potential-kinetic ratios. The presence of these rings is sometimes seen in atmospheric layering. Field speeds are the defining codes of these rings which have mathematically precise boundary divisions. The concentric rings of the GST Field describe a specific environment to which materials will relate, in which they may float.

With the application of heat to a material the kinetic energy is raised and its buoyancy ratio is, therefore, changed. This illustrated by hot air balloons. As a pilot puts on his jets and heats the inert gas in his balloon, his big colorful balloon will rise up in the air. When the pilot wishes to descend to the ground, he lets his balloon cool. He can float about over the countryside as the warmth of his gas balloon is just right to balance the temperature of the currents.

When ocean water is heated it will bulge upward. When water boils it become gaseous and rises upward. When molten brass cools it shrinks. Heat causes expansion and also increases buoyancy.

Gasses are the most buoyant materials. They are kinetically active and freely move about, but they will layer themselves. Hydrogen is your lightest gas and quickly falls upward. Helium will float in the atmosphere whereas carbon dioxide tends to sink. Fluids represent the next octave range of S field buoyancy. Within the fluid groups, pure water is used as a standard. Salt water will float above pure water while oil will float atop both.

Harder materials can be measured for buoyancy in two ways: (1) how they float in different fluids; and, (2) how they layer themselves when heated to a fluid state (and their melting point temperatures). When heated by volcanism, rocks will layer themselves in strata. Basalt rock sinks down, granite and silicon aggregates rise up. Iron sinks down, copper rises up.

There is an enormous amount of S Field packed into the heavy element of uranium. That S Field is compressed and quiet. It does not become active until the atomic structure is knocked apart. When that happens S Field explodes outward and upward and you call it a bomb. Both weight and buoyancy are relative measures within their environmental ST Field conditions.

The sun's light upon earth can change the gravitational speed in a given area. It interacts with the GS Force rising out of the earth. Together they alter an atmospheric condition you measure as barometric pressure. A "high" barometric reading indicates a GS Field expansion. A "low" barometric pressure indicates a GT Field predominant atmosphere. These conditions have a slight effect on weight and buoyancy. This effect is particularly evident in plants and animals. A "low" pressure subdues biological activity while a "high" barometric pressure energizes growth and activity in plants and animals, and of course, human beings.

SPACE FIELD IS POWER - THE GS FIELD

Every action in a T Field predominant sphere is balanced by an equal and opposite action in the S Field predominant sphere and visa versa. Although it is repetitious to make this statement often, it serves to remind you of the ever present power of the S Field. We often speak of electric action in old conventional ways, in shortcut explanations. But it is important to remember that S Field drives the process of electromagnetic activity. S Field is the sphere that defines the circuit loops, the rhythms, the wave amplitudes. Each electron center, each spark is defined by radiating S Field.

The S Field drives the rotors on your electric motor. S Field can be drawn off and stored under compression. It can be used as a condenser. S Field is everywhere around you and can be gathered up just as you need it.

S Field is the main attribute of resistors. S Field sheaths around a wire, used as an insulator, gets your T Field electron to the place where you want it. It was the S Field energy you saw in the mushroom cloud of the atomic bomb test. It is the rapid release of S Field that creates a dynamic explosion. When you are looking for power in action, you are talking about S Field. S Field pushes T Field and makes it move.

In the future you will discover how to separate fields in a controlled way. S field will be used to levitate vehicles in gravitational fields. Concentrated S Field, moving at controlled speeds, can surround a heavy object, break its GT Field linear connections, and lift it upward. Gravity, which has been such a problem for you, can be turned into an asset. succumbs to the downward pull of gravity and is grounded. S Field pushes. T Field pulls. S Field expands. T Field CLOSES. A rocket that has S Field thrusting it in stages can travel much father than a bullet. S Field push or thrust has been the primary interest of your engineers since the days of "cave men." It remains necessary to know much more about how the S Field operates as a force. To understand the GS Field we will have to unders tand activities of the S Field.

CENTRIFUGAL FORCE - THE SPACE TORUS

Any action in one field solicits an equal action in the opposite field.

A particle is what you call an organized unit of T Field predominant energy. Therefore, a traveling particle solicits a response from the S Field of the general ST Field through which it travels. Let us use a traveling bullet as an example. See **Illustration 3-7**.

The nose of a bullet penetrates a field at a specific directional speed. The field is shocked at its presence and reacts with equal and opposite strength. The penetrating tip of the bullet meets a sphere of opposing S Field. Astride the tip the S Field coils in the opposite direction to the travel of the bullet. It coils in a tight torus around the tip, the walls and any traveling T Field point. The coiling torus then begins to expand its diameter outward, perpendicular to the direction of travel. As the bullet passes by, the S Field continues its outward expansion until it is dissipated. The donut shaped torus expands as time is expressed. (You must notice that the torus has electrical properties. It is activating and altering the environment. Any motion in air is the result of the field action.)

Every moving action will show a field shock wave with kinetic response equal and opposite to the force carried by the moving particle. The shock wave is made up of expanding, rolling tori that INSULATES the traveling object. The faster the travel, the greater is the insulation of the S Field. An S Field is a hard wall to a T Field.

> **The torus around a traveling particle insulates it and prohibits its movement from side to side. This concept has been called centrifugal force. It is operative in any special environment.**

This concept is worthy of your careful attention, so we will create a picture for you of how it works. Supposing you took fifteen automobile inner tubes, inflated them, and glued them together in a line, one after the other. They would sit on the grass looking like a giant caterpillar. Now supposing you tried to crawl through the center of these tubes, end to end, but halfway through the tubes you decided to make a turn to the side. Could you make a turn? The only way you could turn is if one side of a series of inner tubes was deflated and the other side was inflated. That would take a lot of force and outside help. Inside the series of tubes you would inevitably choose to continue your forward path. That is the situation of a T Field particle traveling with the baggage of a space torus. This principle is based upon the law of the equal and opposite action of the ST Fields. See **Illustration 3-8**.

ILLUSTRATION 3-7

LAW: ANY TRAVELING BODY STIMULATES AN EQUAL AND OPPOSITE FIELD AROUND IT.

ANY TRAVELING MATERIAL BODY IS T FIELD IN MOTION. EVERY T FIELD POINT AND EVERY OVERTONED COLLECTION OF T FIELD POINTS DEVELOP S FIELD TORI SURROUNDING THAT MOTION

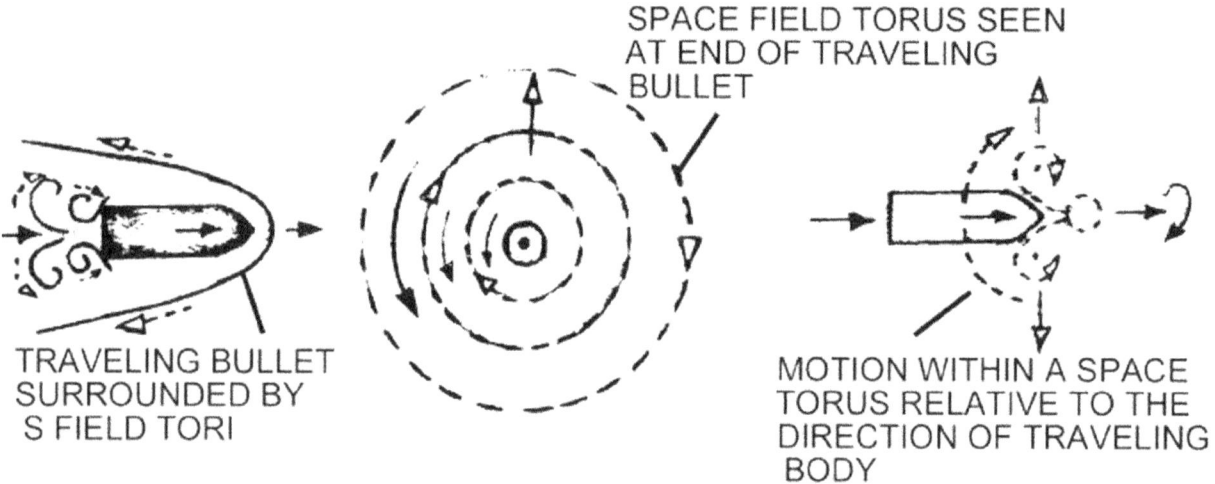

SPACE FIELD TORUS SEEN AT END OF TRAVELING BULLET

TRAVELING BULLET SURROUNDED BY S FIELD TORI

MOTION WITHIN A SPACE TORUS RELATIVE TO THE DIRECTION OF TRAVELING BODY

TRAVELING BULLETS STIMULATES S FIELD TORI IN SPACE-TIME INCREMENTS. AS TORI EXPAND PERPENDICULAR TO THE TRAVEL, TORI DISCRIBE A WAVEFRONT

THE CURRENT TORI DIAMETER IS SMALLEST, THE OLDEST TORI HAS THE BIGGEST DIAMETER ON A PLANE PERPENDICULAR TO THE TRAVEL OF THE BODY.

THE MECHANICAL POWER OF THESE TORI ARE IN DIRECT RELATIONSHIP TO THE SPEED AND MASS OF THE TRAVELING BODY.

ILLUSTRATION 3-8

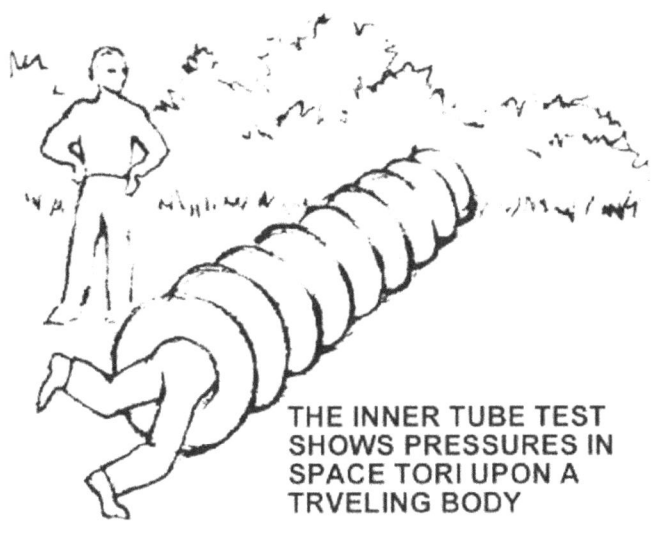

THE INNER TUBE TEST SHOWS PRESSURES IN SPACE TORI UPON A TRVELING BODY

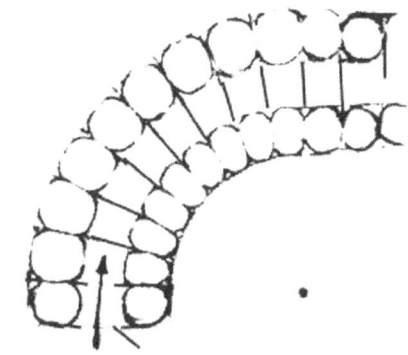

WHEN THE INNER TUBES MAKE A TURN IT IS NECESSARY THAT PRESSURES SHIFT WITHIN THEM.
THE DRAWING SHOWS A CUTAWAY VIEW OF INNER TUBES

TORI HAVE A LOT IN COMMON WITH BALLOONS,
EXCEPT THAT THEY ARE USUALLY
SHAPED LIKE AN INNER TUBE FOR A CAR
SURROUNDING AN ACTION.
THEY HAVE INTERNAL PRESSURES
THAT RESIST CHANGE IN FORCE DISTRIBUTION.
THAT HAPPENS BECAUSE OF THE NATURE OF RADII
IN AN S FIELD SPHERE.

It has been reckoned that a spinning object exhibits a force that pulls out and away from the direction of the spin; that any object will travel in a straight line unless acted upon by an outside force. That has been called centrifugal force.

For those students who would require experimental verification of space tori, we refer you to the work of Heinrich Lentz. Lentz's law describes that when a magnet moves through a wire loop, it develops a current in that loop which is opposed to the direction of the magnet's current influx.

In a quiescent field a traveling object will continue a straight course. In a traveling field, such as a gravitational field, the object and its torus will move with the field, unless the object's directional force is greater than the field force. The paths of traveling fields also show tori. Tori surround each gravitational T Field line, separating one from the other.

When an object travels rapidly and laterally to the GT Field lines, it is insulated with space tori. Those tori discourage the penetration of GT Field lines into the interior of the object. The straight line path maintained by the tori may completely overcome the directional thrust of the GT Field lines. All aircraft are functional because of this principle.

In the design of a "UFO," the craft is also insulated by S Field tori. But those tori are not the result of speed but are produced by motors that issue rolling tori continually out and around the ship to regulate any T Field line penetration. The speed and vibrational rates of those tori must be matched to the condition of the GST Fields of a planet. Your engineers are working on mechanisms to separate fields but without knowledge and proper elements their attempts will continue to be clumsy.

In conditions of thrust when a space torus surrounds a traveling object, it is the space torus that acts to penetrate a resisting object. Kinetic force opposes T Field bonds. The winner in a collision has the most forward energy per smallest area of surface. It may appear like a hard bullet penetrates a brick wall, but it is really the space torus around the bullet that does the penetrating. The space that breaks the brick is the space that was inside the brick as well as the S Field torus at the tip of the hard bullet. (A space torus does not travel with the object, but is created with each increment of field penetration.)

When a potato falls onto a spring scale, or when a falling object hits the ground, you are describing conditions of thrust from an S Field torus. The force of that thrust is due to its kinetic expression in cooperation with the T Field points of that which is moving (a field or an object). S Field is a hard wall to T Field. T Field is a hard wall to S Field. S Field flows with S Field. T Field flows with T Field in bonds. And, no field force can be considered separately from its environmental fields.

This challenges the idea of collected "mass" as a singular force. "Mass" generally refers to quantities of T Field points bonded in a unit. That condition results in an equal and opposite presence of space tori under pressure. (Do not forget that earth and its materials are always traveling at high speeds through the cosmos.) When speed is added to an object, even in small increments, the character of force is shifted toward the kinetic side of an equation. T Field bonds are loosened, expansion occurs. Space tori surround EACH T Field point in the object (as well as its outside form). This S Field is drawn from the environment and spins in opposition to the direction of travel. The ring of the torus moves outward perpendicular to the direction of travel. T Field is also drawn from the field at-large to balance the kinetic expression. Universal fields stabilize the equation.

It has been observed that once a traveling object, such as a rocket breaks away from the edges of gravitational pull into outer space, the speed it carries at that point in time is maintained as the rocket travels further into spaciousness. That constant speed is sustained by the precise balance of the fields. Even though the propelling thrust is no longer behind the rocket, the speed does not change. The T Field potential supports the S Field tori; the space tori supports the integrity of the T Field potential. The Universal Fields provide the additional energy needed to fulfill the action. The direct relationship between the S Field and the T Field is sustainable as perpetual motion as long as no outside force interferes with the action. The object is said to have momentum in an inertial field. But it is more accurate to say that an object's speed and T Field are perpetually balanced by its space tori and its Universal fulcrum.

Your common experience with space tori is when you are riding fast in a car and turn a corner quickly. Your body wants to go straight and you have to put up some strong resistance to pull your body around into the turn and the new direction of travel. That old torus keeps going in a straight line. Then you can relax again as your new space tori continue to do what they always do, to keep you traveling in a straight line. The tug on your body had little to do with air, only with equalizing pressures in the field that are around you all the time.

CENTRIPETAL FORCE

There is only one true centripetal force, that of CLOSURE. The effect of CLOSURE is to potentially strengthen the center of a sphere. Centripetal force can be enhanced by applying a force of pressure to an area in order to slow it down, in order to favor time of duration over spacious travel in its equations. Centripetal force is characteristic of the T Field. It is not a correlate to centrifugal force.

PRESSURE AND HEAT

Pressure of CLOSURE is comfortable to the T Field as it fulfills its desire for duration. Pressure is resisted by the S Field in its need for expansion. Pressure causes S Field to circle tightly "in place" instead of swirling outward. Forces of pressure come from T Fields establishing bonded limits. As pressure is increased, so is potential. S Field surrounding that increased potential will spin around in a smaller circle more times per second. Any radiative effects show smaller wavelength and high frequency. Among those radiative emission is a wavelength recognized as heat.

Wavelengths of heat appear on the electromagnetic scale in the infrared zone. While it is S Field predominant it demonstrates certain T Field attributes in that it is quantifiable. Like visible light, heat is signaled into a field by agitation from friction and pressure. When you study about light you will gain a clearer picture of heat resulting from resistance.

The construction of a refrigeration appliance begins with a gas under pressure in a metal chamber. The gas in resistance sends field signals outward in the form of heat through the metal chamber. Once the gas is "cold" it is sent through tubes in the area to be refrigerated. The tubes pick up and transfer the heat to the gas within the tubes. That allow the gas to expand. The expansion of the gas takes up heat from the area and is returned to the compression chamber where the "heat" is to be once again removed from the gas. This procedure sounds as though "something" was picked up here and deposited there. Instead, it is a situation of stop and start.

The principle of heat transfer is based on the greater principle of equilibrium within a field. Realizing that ST Fields move freely in and out through material such as metals, you know that those fields carry vibrational signals. Signals in the heat zones cause field molecules to vibrate and move about, distorting the fields at large. (The high heat from a stove causes such agitation in fields as to deform light signals coming through it.) The Law of Equivalencies in fields effects a diffusion to balance. What seems to be substance is instead a mathematical ratio of field speed. The Law of Equivalencies will be discussed in Chapter 4. It is a Law adjudicated by the greater Universal Fields and while its effects are pervasive, very little can be understood concerning its Divine origins and details of its interventions.

Because of signal transfers in a field and effects of diffusion you have the phenomena of fire burning, one corpuscle exciting the next adjacent corpuscle to a point of ignition. Heat signals break bonds in T field structures. While the expansion of S Field will take the heat signals up and away.

If a student of religions were to examine the ancient Hindu texts and attempt to grasp the aspects of the god-goddess Siva in terms of the laws of Symmetric Field physics he/she would find that mankind was knowledgeable from the beginning of civilizations.

Heat, like all Space and Time Fields is both signal and substance. S Field in vibrations of heat breaks T Field bonds of substance, while in another vibrational state nurtures the formation of bonds in spherical cocoons.

Warm air rises in the GS Field. Expansion of the S Field will always stop the vibrations of heat. Cold air drops as its slower speeds form an alliance with the GT Field equations. Cold air at the surface of the earth is once again under pressure and is agitated by the friction of the planet to start a warm vibration. Again the air rises and circulates.

The microwaves in your ovens are not hot. Their vibrations agitate the water molecule into action and internal friction starting a generation of heat in your food. Your studies need to focus on the relationships of ST Field vibration to motion, signal to motive force, and its precise management.

Mankind is again jolted from his blissful ignorance in his earthly Garden of Eden. His nudity is exposed as he engages with the scope of the feminine S Field. His intelligence and wisdom alone will carry him safely through the coming millennium. He and she have partaken of the tree of knowledge, provided by the feathered snake who takes his own tail in his mouth, the cosmic circuitry.

MOMENTUM

Behind a traveling object, such as a bullet, is a remnant of the S Field wavefront thrust from the gun powder. As the bullet travels with its space tori encircling its trail, a T Field current is generated behind the bullet moving in the same directions as the bullet. This T Field current helps to sustain the bullet's travel. It strengthens the thrust of the wavefront. It provides an enhancing field resonance for the action. The S Field tori expand behind the bullet, actually pushing it forward. The bullet exhibits extra force even after the force of the gun powder has ceased, as a result of the squeeze from behind from the space tori. This extra propulsion is called MOMENTUM.

A frequent experience you have with space tori is when you take your motorboat to the lake. The waves you make in the water are not only from your bow dividing water but from the tori shock waves that develop in the fields and air and water encircling the boat. **Illustration 3-9** will show how the space tori create the ever expanding water waves behind your boat. And when you shut down your motor the boat will coast along for a minute or two. You have to know just when to shut off your motor as you approach the landing dock so that you do not strike the dock. You have learned about MOMENTUM.

When you drive in your truck carrying a load of brick you know that you will not want to make any quick stops. If you do your load could slide forward on the truck bed, with disastrous results. The heavier the load the harder it is to stop its motion. The heavier the load the more powerful is its space tori.

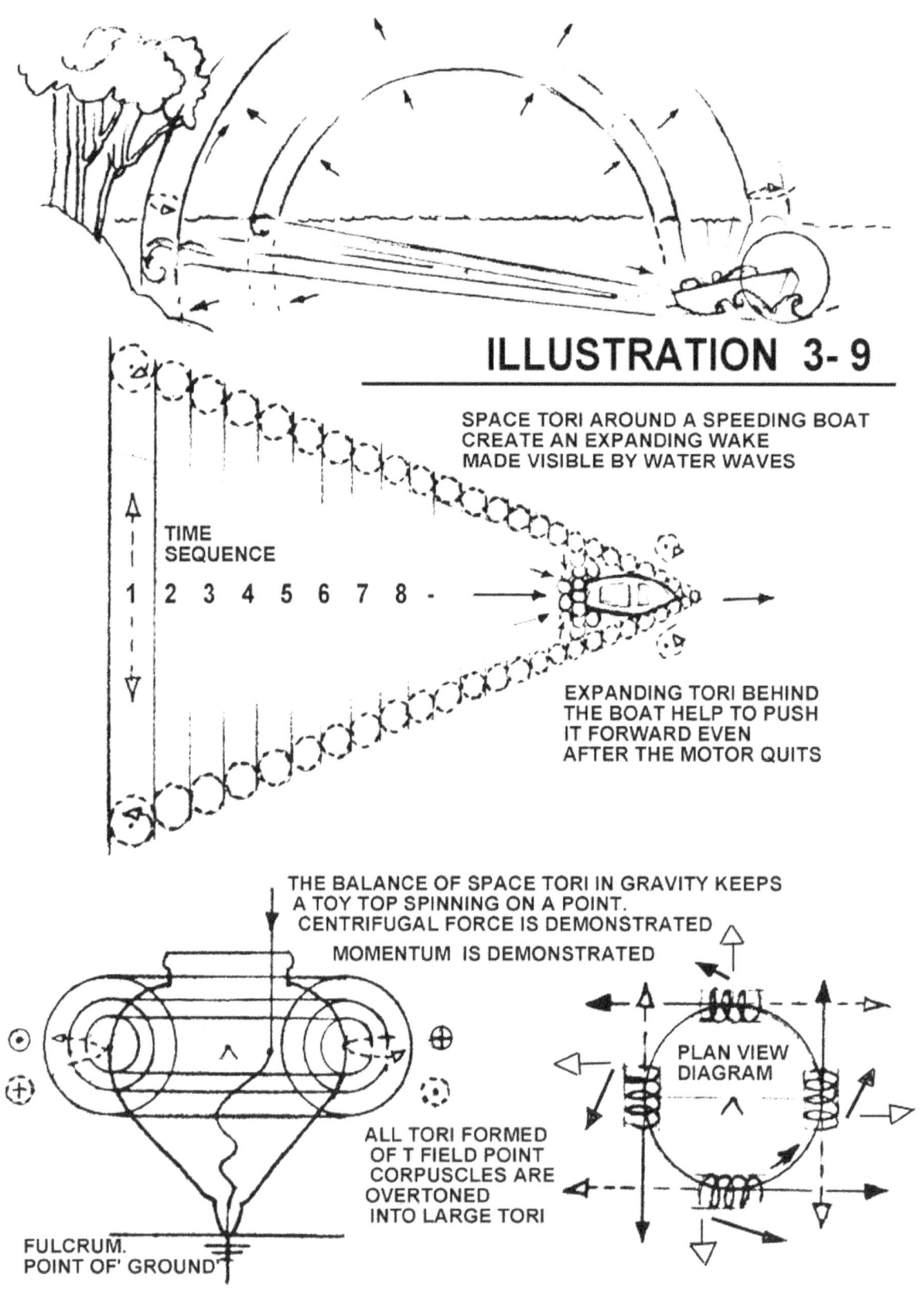

Now you know that **MOMENTUM is the result of space tori, S Field in motion acting to complement T Field.**

INERTIA

A heavy object is still on a table (still, relative to its field in motion). Each quiet T Field point is threaded with a GT Field line which continues its slow path to the ground. If the object is moved, its motion causes its T Field points to develop space tori around each one. the kinetic tori breaks the connection of a GT Field line to the ground, in a gradient relative to its speed. Any GT Field line, especially a slow moving one, is resistant to being broken. The line has some elasticity, but it is tenacious in holding together in CLOSURE. A little lateral kinetic force will bend the GT Field line while a lot of lateral kinetic force will break it off. There are many lines to break to get the object moving relative to the table. The GT Field lines resist motion in opposition to its field flow. (This resistance can be measured using a pendulum.) A heavy object on a table has additional friction to overcome. It takes more force to get an object to move than to keep it moving. This resistance is called INERTIA. INERTIA refers to the force necessary to overcome, in order to break T Field lines of CLOSURE and to put a space torus into action in a new direction and speed.

When force begins to motivate S Field into active tori there is some resistance present. In addition, with motion, the USUT Field engages in the object's speed. That takes time. Motion to change speed and direction is rather slow in a gravitational field and seems to be a "decision made by a committee."

The principles of INERTIA, MOMENTUM, and CENTRIFUGAL FORCE are all evident in the action of a fly wheel. And it is CENTRIPETAL FORCE of CLOSURE that holds a fly wheel together in all of its S Field stresses. Can you now write a paper on flywheels and reevaluate your formula concerning "angular momentum"? Can you understand the field forces present in the action of a gyroscope?

A strong point to be made is that while an S Field tori encompas ses a whole object, it also encompasses a singular T Field point within that object. The whole of an object in space will try to travel at an unresisted steady speed as does each of the parts of that object. If an object enters space travel tumbling, it may continue its tumble as each part tries to maintain its own speed. The spherical integral forces will try to bring the incompatible and irregular speeds into balance over time. An object tossed into space with a spin will eventually stop spinning. The rotation of a plane, however, is due to circuit loop flow, while a simple tipped axis may be perpetuated by space tori. Still, a planet travels the same speed as its orbital speed, having a comparatively slight space tori. If you can recall the position of the solar system within its galactic orbital flow, you can see that no single planet (on the equatorial plane) and no single point on those revolving planets is fluctuating with regard to its forward flow. (See **Illustration 2-6**).

Cosmic bodies with oblique orbits (such as moons) are under the influence of a lower octave force. Comet are traveling bodies that break away from spinning systems with strong space tori. While trying to cut a straight line, they cross into circuit systems that alter their courses. The strength of a comet's space tori prevents it from becoming trapped in a solar system.

Momentum is a word that describes a perfect balance of space field tori around a T Field point in motion. That perfect balance sustains the kinetic performance. The word inertia means maintenance of existing balance and does not really qualify to be called a principle in itself. Inertia occurs when any change of motion occurs. Inertia acts against the force that demands that change. Inertia is only an extension of the principle of MOMENTUM which is part of the LAW of EQUILIBRIUM.

TIME SPENT IS MONEY EARNED

You have an interesting way of speaking about time. You say that you spend time, and save time, waste time, and use time wisely. It is as though you could think of time like money. "Time is money," you quip. If you think of potential as having money stored in a bank vault, you can think of time quietly stored in a center. As you spend money you can spend time on activity. The force of gravity translates time into active space field starting at "point of change=0." It goes out and away as though it was gone forever. You can say that the time-money was spent. Or, you can say that the time-money was invested and that you expect to get it back with "interest." As space field actively circulates, it also creates and nurtures new energy. As it reaches a circumference it touches its "point of change=0," and turns again into time-money, then returns to you with "interest." Money and time must be circulated in order to sponsor growth. And it must cyclically return to rest in its storage bank. The process of circulation and return can be mathematically monitored. Its harmonic activities can be understood clearly. Physics, like money, can be a part of everyone's life and can be clearly understood.

CHAPTER 4

A NEW WAY TO SEE

RADII - AN OUTSTANDING PHENOMENA

Energy always takes the form of a sphere with a center. Radii are straight paths between the circumference and a center of a sphere. Radii come in pairs by definition. One directional radius balances another opposite to it. Together a pair of radii become a diameter. Energy travels on these radii, either out from center or into center. Radii do not curve unless a center spins at a different rate than a circumference. Radii often branch like a tree to accommodate unusual configurations.

The radii lines are an elastic and contractive force of the T Field. These lines have length and no width; they are one dimensional. Their length attempts to described the potential at their center. Two radii together locate three points in space which define a plane. Radii lines become dimensional locators for dimensionless points in space.

Radii are paths that transit T Field energy from the circumference of a sphere to its center. But radii are always present, just as all axes are ever present. There is an old joke about a stranger who approaches a farmer fixing his fence. The stranger points to a dirt road and says, "Old man, can you tell me where this road goes to?" The farmer replies wryly, "Well sir, that road has stayed right there for as long as I can recall."

Radii are like roads, always there as outreaching paths in a field. They are partnered by the radiant S Field which encircles and sheaths each line. Because of radii we have visible light reaching from the stars. No matter where you stand you can see the same star without dark places in between. No matter where your big telescopes travel in space, they can focus on a very distant star. How many radii can the energy of one star provide? Too many to count. Because of radii you have S Field reaching explosive dimensions in spherical shapes. Radii are like the bones of an organism that we call the Symmetric Fields.

SPACE-TIME FIELDS MEAN LIFE

Probably the most important function of the ST fields is to provide systems for life energy to express. Whatever applies to the ST Fields as functional law is also the lawful premise for biological designs. As you read and understand the ST Field systems, think about primary cell creation and cell division in all biological development. Think about how plants shape themselves and process food, and exchange energy with the earth and sky. The stuff you know as matter in living forms is made fro m ST Fields. The materialization comes about well after the invisible fields are properly organized as ideas, in signals. When you observe a cell, or a DNA helical structure within a cell, you are seeing less than its whole. These small but complicated parts have embedded within them volumes of whole systems that do not occupy space as you know it. The systems are miraculously reliable. Parents, seeing their newborn baby for the first time, experience fully the wonders of the cosmos with a thrilling sense of awe. No science can consider less than that.

When we are talking about signals, pressures, volumes, waves and forces we are laying the foundation of knowledge upon which your biology and ours is built. Be patient to slowly draw this knowledge deep within your being as it will serve you in futures that are yet beyond your dreams.

SPACE FIELD AND UNIVERSAL FORCE

We have talked about how a falling body gathers T Field into itself as it balances its speed with "mass." Now we wish to point out that the S Field sphere must gather more S Field into its form as the bonding of CLOSURE intensifies a central potential point you call "mass." S Field increases its volume and/or its pressure as the stillness of potential translates into extended duration. When new S Field and new pressures are required a sphere calls upon both S Field and US Field to compensate a loss of balance.

A mystery remains for us to unravel. We experience that any sphere under pressure will register EQUAL PRESSURE at any place upon or within the sphere. Pressures within are balanced by appropriate surface tension, and that tension measures equally at all places on the circumference. We know that spheres become larger or smaller as center potentials become greater or lesser. The expansion of spheres happens like growth from the inside toward the outside. Concentric rings called waves start to move outward as the sphere expands. A sphere in contraction reverses the process. A sphere at rest, however, shows stable concentric rings into which atoms can settle. Radii define the relative positions of rings. Radii centralize each equal volume corpuscle.

The force of S Field behaves differently than T Field. S Field expansion and compression **multiply** (not just a shortcut for addition) under LAW. This mysterious LAW begins at the beginning of God's design of fields and cannot be further explained. With equal and opposite precision the T Field expands and contracts by digital addition and subtraction.

Any hydraulic lift used in industry is based on formulas of pressure that prove to be consistently true. **Illustration 4-1** shows an incompressible fluid confined under pressure in two connecting tanks. Piston A supplies pressure to the fluid at 10 lbs. per square inch with a volume displacement of 100 cubic inches. The pressure is transferred throughout the fluid, measurable at any point at 10 lbs. per square inch. The piston B has a much broader diameter covering 1,000 square inches. This piston is designed to lift heavy weights, such as trucks in repair. The force of pressure is directly transferred in multiples from piston A to piston B by the connecting fluid so that the 10 lbs. on piston A now registers on piston B as 10 x 1,000 = 10,000 lbs. The amount of fluid displacement by piston A, 100 cubic inches, will be the same volume added to piston B. Since piston B has a base of 1,000 square inches, the piston will be lifted up 100 divided by 1,000 or .10 cubic inches. The force of pressure is multiplied but the volume stays in direct ratio.

ILLUSTRATION 4-1

PASCAL'S PRINCIPLE OF HYDROSTATICS

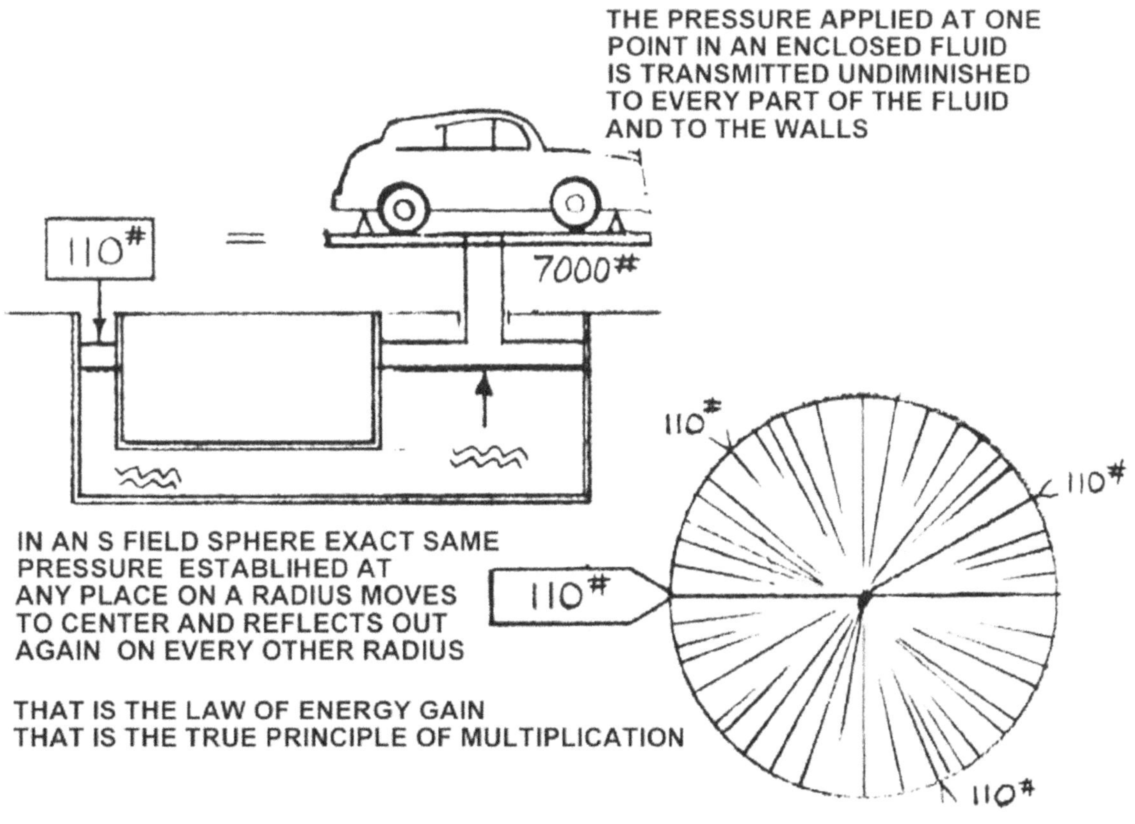

THE PRESSURE APPLIED AT ONE POINT IN AN ENCLOSED FLUID IS TRANSMITTED UNDIMINISHED TO EVERY PART OF THE FLUID AND TO THE WALLS

IN AN S FIELD SPHERE EXACT SAME PRESSURE ESTABLIHED AT ANY PLACE ON A RADIUS MOVES TO CENTER AND REFLECTS OUT AGAIN ON EVERY OTHER RADIUS

THAT IS THE LAW OF ENERGY GAIN
THAT IS THE TRUE PRINCIPLE OF MULTIPLICATION

Your automobile has a hydraulic braking system. When you put 10 lbs. of pressure onto your brake pedal, the pressure is multiplied onto the brake drums of the wheels. You are able to stop a 2,000 lb. car traveling 50 miles per hour with your toes. Where does all that energy come from?

To begin to understand that, you have to ask yourself how much energy is stored in a dimensionless T Field point in DURATION. The answer is unfathomable. It is enough to know that it can be described as eternal. When a measurable amount of energy is compressed against one radius path the message is taken to the T Field center point. From that point all the other spherical radii are informed with the identical message. We call that message a signal. The signal then includes a spatial-mechanical force aspect. Every radius becomes forceful. Since radii can be any number (n) the force is multiplied any number of times. Signals transform energy. Signals create energy.

S Field, coiled around the T Field radius, compresses like a spring. Overtones are compressed and multiplied as overtones. When more energy is added to a sphere its radii normally grow longer, the sphere becomes larger. When a sphere is under compression and cannot grow larger, it becomes corpuscular, it divides. Each new division carries the same signal of forceful pressure within each part, (each corpuscle) (each crystalline unit) (each cell). Replication takes place in less than a blink of an eye. It is the nature of the S field to nurture and give birth to new energy according to the signals implanted and reflected within it. We can refer to this activity as SPHERICAL ENERGY GAIN. The universes today are expanding in a condition of mathematical powers. The universes began in a condition of mathematical roots.

The principles of SPHERICAL ENERGY GAIN is most apparent in fluids and gasses, but it is always present in fields, in solids, in atoms and in cosmic formations. Many mechanical inventions that necessitate counterbalancing strong forces, such as fork lifts, auto lifts, jacks, brakes, shocks and pumps are engineered to use this principle to great advantage. Gasses such as hydrogen, helium and air are used to lift weight bearing balloons, lift rockets, and give your car a smooth ride on rubber tires.

Fields are always in compression when matter is formed by replication. The constricting nature of CLOSURE in bonding puts S Field into contained pressure by the action of strong surface tensions. When bonds of CLOSURE break apart the S Field is released, often with explosive force. If kinetic activity is added to S Field in compression, the S Field can break the bonds that hold it. That happens when any fire burns. Heat from one molecule activates heat in its neighbor so that it, too, ignites (breaks is bonds). That happens as your cook stove brings water to a boil and S field turns fluid to gas. That happens in your steel mills as furnaces bring metal ores to a red melting point and liquid metal flows into ingots.

It should not have surprised you that strong concentrations of T Field, such as you have in atomic nuclei, are surrounded by strongly compressed S Fields. And that when atomic bonds are broken, S Fields are suddenly released in massive explosions. That simple discovery changed the politics of your world with atomic and nuclear bombs. Your next scientific challenge is to separate helium into pure S and T Field components and bypass the explosive stage.

It is possible that energetic fuels can be produced biologically in sufficient quantities if your consumer desires could be held in check. Consider using a plant like algae. Your ocean water is your largest resource. Winds and ocean currents and tides can be utilized to generate electricity. SPHERICAL ENERGY GAIN is available all around the globe. It can become available in one engineering leap.

The resources that the Star People use come from very dense planets not favorable to life. They are used for mining only. The minerals have a high level of radioactivity which is used directly to generate a power greater than your electricity. Yet we have little need for its use. Our biology is more advanced and serves us well.

Some energy is stored, some energy is created anew as needed. The ideas that signal the formulation of energy are usually not new, but systematically prepared to perpetuate an abstraction. Energy is often used again and again in different organizations. But it is the ST Fields that multiply the universes with new energy. From the ideas of the center IAM added to the spherical IAM powers come the expanding reality of Universes. Those ideas are coded signals that have no dimensions but find expression in the geometric rhythms of dimensionality. New energy springs from the whole of IAM.

Energy is often borrowed from the storehouse of undifferentiated fields. Any emerging events in the ST Fields automatically solicits a loving support of the USUT Fields to contribute equal energy and balance.

MR. BAKER'S EQUAL-VOLUME COOKIES

To understand your physical world, which is full of strange dimensions and forces, we want you to take an imaginary excursion to a bakery shop where Mr. Baker makes very unusual cookies. This may sound like a kid's game to you, but you will find there are many occasions to visualize Mr. Baker's cookies.

At his shop, Mr. Baker, the cookie baker, presents to you a dozen round balls of cookie dough, all looking alike, all of equal volume. Now it is up to you to tell him what shape you would like your cookies to be. They can be a long roll like a primary axis. They can be flattened like an equatorial axis. They can be patted into a thin pancake or made into a stubby rectangle.

Then Mr. Baker shows you how he can pack his cookie dough tightly into his big bowl. He starts by patting each equal-volume cookie around the inside of his bowl. The he adds another layer exactly on top of the cookies of his first layer, then another layer and another, until the bowl is almost full and flat across the top. Each layer is a little fatter and a little smaller in major diameter. His final cookie is placed in the middle patted into a half-round. All the bowl space is filled. All cookies are interconnected in a radial pattern. This is how Mr. Baker can store his cookies in the refrigerator until he is ready to bake. It is important that each cookie will have equal pressure inside of it. In that way there will be harmony in his little shop. **Illustration 4-2** will show you his method. For you to understand wavefronts and signals in space, we hope to have provided this comprehensive picture of spherical organization.

Because of the LAW of SPHERICAL ENERGY GAIN all parts of a sphere equal any number necessary to supply equal volume and equal pressure. All divisions of equal volume must also satisfy the influences of harmonies and speed. If you take away a cookie from the bowl of cosmic spheres it will be instantaneously replaced. You will always have equal volume multiplied by any appropriate number (n).

Looking once more at the deformed cookies (which represent deformed and polarized spheres) you will notice that there is a changing relationship between axes x and y and z. When pressure is applied to shorten axis x, axes y and z become longer. A severely deformed volume will protest and all will divide again to distribute pressure as equally as can be.

ILLUSTRATION 4-2

PRESSURE AND POLARIZATION MAKE MANY DIFFERENT SHAPES IN THE CORPUSCULAR FIELDS.

BECAUSE OF THE LAW OF ENERGY GAIN CORPUSCLES WITHIN AN OCTAVE GROUP MAY DIVIDE AND RESHAPE ALL AT ONCE TO MAINTAIN EQUAL VOLUME AND PRESSURES WHILE THEIR SHAPES MAY VARY WIDELY

LOOK INSIDE A MOLECULE

In E. W. Müeller's marvelous photographs you can see how the harmonic pressures of one sphere balances with the harmonic pressures of another. Pressures and spaces interrelate in symphonic agreements to orchestrate an element as precious as gold. In this picture a white dot represents at least one atom. The photographs were made by bouncing ions off carefully sliced metal. This work was pioneered by Dr. Müeller and is known as Field Ion Microscopy. He has given mankind a way to look inside the structure of a molecule. **Illustration 4-3** is pure gold. Detailed discussions of the photographs will be presented in Chapter 11.

You will see that the atoms arrange themselves according to patterns in the ST Fields (which appear black). Fields are invisible except as tracks into which atoms fall. Fields show up as footprints mark the sand. But it is fields that create the atoms and orchestrate their activity. Each dark circle center in the photo is an S Field circuit with a T Field center. Concentric wave rings alternate between S Field predominance and T Field predominance. The atoms settle themselves in the T Field predominant rings (orbits). Notice that the intersection of T Field rings provides a stable potential place for an atom to reside. These photographs clearly show the positions of radial axes that are an automatic part of every sphere (they show as lines in concentrations of white atoms).

On close examination, the photo reveals a wealth of knowledge about fields and materials. Chapter 11 will tell about the details of mineral crystallization. Notice here that the orbital rings closest to the dark centers carry greater potential, transit a shorter distance, and exhibit wider S Field sheaths. Notice the harmonic geometric patterns based on divisions of a sphere. Especially notice that the black (invisible) power centers hold atoms in subordination. Substance is the result of bonded ST Fields, not atoms alone.

Rays of light signaled from the sun have wavefronts many times larger than an atom. Light cannot be reflected from an individual atom. Ions are smaller than an atom and can be used to bounce off an atom and return to a camera, thus making an image of the position of an atom as the strike emits a "photon" signal. Many ions are absorbed into fields that the camera sees as dark. The rings of ST field spheres are too fine to reflect either sunlight or ions. They are detected as shadows in between, and sometimes by finely tuned electronic instrumentation. Be assured they are there. Vibrations in the ST Field can be detected by the human nervous system, but seldom understood.

Science has found ways to amplify signals within molecules and atoms. By amplification they can make careful use of those signals and programs of construction. That knowledge is now being applied in the computer sciences. Controlling ST Fields is to control everything mechanical, electronic and biologic.

ILLUSTRATION 4 - 3

Dr. Edwin Mueller's famous photograph of crystaline gold was taken by his methods of ionmicroscopy. Each light dot represents one atom. All atoms are arranged by the ST Fields into measurable geometric configurations.
Notice how all spheres agreeably intersect one another.

Star people (and others) can identify and record the fine vibrations of field energy emitted from your brain. We can read your thoughts. Star People (and others) can broadcast messages into your brain as sound or light. We are capable of doing this without instruments. Others may use electronic instruments to send messages directly into your brain. You will hear it and see it. Do not mistake it for your own thoughts. We will identify ourselves. It is "others" who might try to deceive you to make you think that the broadcast messages are actually your own thoughts.

EQUIVALENCIES

Balances are part of LAW regulated by the agencies of MIND. Justification is the work of the Universal Fields as they engage the activity of the ST Fie lds. We can explain what balances do, but we cannot explain what balances are. Because of the nature of balance and its cosmic integration, the ST Fields can find a safe expression in many variations.

Because of Universal balance, time and space can express in unity as speed and harmonic variation, there can be S Field predominance and T Field predominance. Instability can have creative expression without disaster. The USUT Fields hold the ST Fields in delicate balance. The function of balance in the USUT Field is very much like your words "Amazing Grace." It is your lifeline that prevents disaster.

US Field also expresses the LAW of SPHERICAL GAIN. If part of an S Field is torn away from a complete sphere it will be replaced in an instant. And the US Field goes one step further. The US Field is there to apply the LAW of RENEWAL. If one part of a sphere is disabled, it will be replaced. If one part of a cycle is disrupted, the cycle will begin again.

> **The LAW of RENEWAL is the Universal power of regeneration, rebirth, adjudication to completion of the DIVINE WILL.**

The LAW of RENEWAL is the tool of the expression of WILL to create, not only reconstruction, but to create adaptive change. There is never an empty rice bowl for it is always filled. If it is not filled in the old way, it will be filled in a new way. Even death can signal a renewal.

When a cat bites the tail off a lizard the entire tail will grow back. When a tree is felled in a forest, another will take its place. When a wolf takes a lamb the grass will grow higher in one corner of the field and set thousands of seeds. As one mountain is eroded to a hill, volcanism will rise up another. The earth itself has times of renewal and adaptive change. No matter what teachers and philosophers may write and preach, the LAW of RENEWAL acts as surely as the sun rises in the morning as a merciful presence the whole day long. Human sacrifice has no effect on this law.

Your personal WILL may evaluate a need for change and adaptation. In that case, although your rice bowl may be full, all may seem insufficient and you may feel wanting. You must then turn to an arbitrator of life. Basic equilibrium can be reestablished, or you may want a change far away from what is normal. If you desire for change is strong enough, a major shift in equivalencies can be mounted and assisted by the LAW of RENEWAL.

The LAW of Renewal is universally applicable. It must be assumed to be present in all studies of physics, chemistry, biology, and in all human or nonhuman endeavors. The LAW is partner to the LAW of Equivalencies. It relates to all shifts in balancing fulcrums and equations. New technologies will emerge from understanding this law.

By shifting your TIME-SPACE fulcrum you can recognize new possibilities of realities in your life, in your present or extended cosmic life. Ideas are invisible abstractions, yet ideas can change your base reality system. Power lies beneath your ordinary senses. Structured power is the creator of your reality. It is yours to use.

Have you ever wanted to fly over the trees like a bird? Would you like to change to a small lightweight body with feathers? What is stopping you? Think about what you could expect from a truly advanced technologically equipped civilization. You are in the planning stage. What is on your drawing board? You can begin to change yourself from the inside outward; and we want to hear from you concerning your imaginative future plans.

Time and space are Universal opposites. One equates to the other as conditions of definition. Space balances on the fulcrum of time. Time balances on the fulcrum of space. The two fields are not at war or in any destructive mode. They are in absolute union, in working agreement. So you can understand that the equivalency between time and space is not abstract but a force best defined as love or union. The Universal LAW OF LOVE is always fulfilled. It is the balancing energy of the ultimate ONE.

We are not the first to tell you that your personal life is eternal. Your energy was designed all new before birth and you will not be swallowed up again by a gray mass of indefinable energy. Although you have brought old DNA to this life and memories from other experiences, you have invented this unique life you have and can carry its essential identity through eons and through galactic spaces. You will not regress in consciousness, but will expand your sensitivities. If this were not so, there would be no reason for us to teach you or take an interest in your welfare. You can look forward to a very long life, like it or not. The rest of your eternal life will benefit from personal and collective planning. You will move on according to the equations of your own design.

If you try to understand Universal LAW, the LAW will work for you and with you. Recognize that LAW is WILL placed as an underlying CAUSE. The LAW of EQUIVALENCIES is the LAW of LOVE. It is the foundation of all other laws and principles and the reality of the equations of physics and alchemy. The LAW of LOVE is the fulcrum of adjudication; it is indivisible, undeniable, certain above all existence and realities. The fulcrum of GOD is conscious above all other consciousness and is all inclusive. The fulcrum of GOD balances all local energy systems and is responsive to the fulfillment of all which is unbalanced. All events of time and space are interlocked within the outstretched arms of the LAW of LOVE. You are in it and of it, interfunctional in its network of life. If you can allow yourself to fully know the reality of love in your total self you can be together with the cosmos in the ecstasy of wholeness.

> **The LAW of LOVE is indeed, the LAW of EQUIVALENCIES**

Your mathematics has made good use of equivalencies in algebraic formulation. What happens on one side of an equation will determine what happens on the other. This is not just a convenience, it is a physical law. We suggest that your mathematics be expanded to include equal and opposite field activity. What happens in the ST Fields will be balanced in the USUT Fields. What happens in an S Field will have an equal and opposite effect in the T field. Each S Field will have an alpha and omega (+, -) and each T Field will have an alpha and omega stage (+, -). Alpha begins as a progressive event at +0. Omega begins as a progressive event at -0. Some of you readers are chosen to develop new systems of mathematics; systems that will include SPHERICAL ENERGY GAIN and the LAW of EQUIALENCIES.

As you pursue your interest in alchemical exploration, you may need to balance and expand your equations of time-space. You will need to connect to the present and the incoming acts of intercommunication on wavelengths that bypass your normal sensory equipment. Do not ask for fulfillment of what you think is missing, as that will be certainly untimely. Ask for a current time construct upon which to balance your equations of consciousness. Then be ready and able to make enormous identity shifts. Be able to die to yourself that you may be reborn to yourself.

We ask you to be a fully realized person in expanded consciousness; for this is the initiation of the alchemist. It is now time for you to lay aside toys and become aware beyond the historic functions of your body and your societies. As we enter this millennium you may set your watches to match our clocks so that we may converse with you. Expand and scan your personal frequency ranges in search of a flood of lighted energy. Your biology will succeed before your S.E.T.I. does. Your heightened amplitude will dance over the tops of mountains. Your outreaching wavelengths will be stretched into overtones that match our distances from you. Listen carefully to what we are saying to you now. There will be a time when you will need to distinguish one voice from another.

EQUIVALENCIES IN THE ST FIELDS

The archer stands poised with his bow and arrow, in love with his target. The **intent** of the archer has already pierced the target as his signal has traveled there, a signal reflecting across an ST Field. His signal has also taken an equal and opposite course behind him. He lets fly the arrow. The moment of release the arrow is on a determined course. **IT IS KNOWN** where the arrow will strike before it reaches the target. Equal and opposite energy follow the course of the arrow striking the target the energy is returned to the archer on the same path. As the arrow hits the target the USUT Field energy that it has accumulated leaves the arrow and continues its path around the world, to meet with its opposing force 12,000 miles away. There are reverberations of the arrows flight on all the radii of the sphere surrounding the archer. The entire event is stored in ST Field centers as accurate history. **Illustration 4-4** graphically describes this common activity.

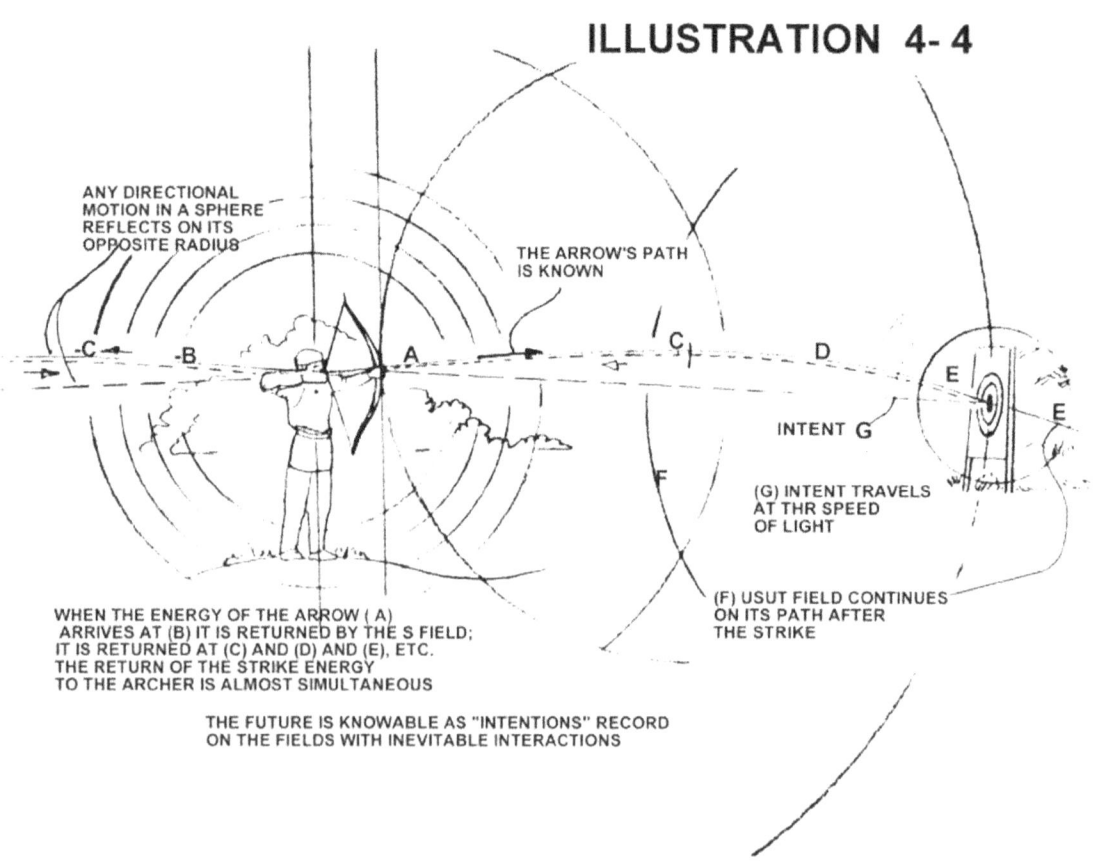

We have described this event of forces as it takes place. We must add that the archer had planned this event days before. Ideas and imagination sent signals to the target days before the event. The target KNEW of the event and agreed upon it. The signals created a "virtual force." If the target had been a man or a deer in full consciousness he would have KNOWN of the event and made some decisions about it. The archer in this event is fully responsible for all the actions and reactions. Only an alert being can cognize the full scope of actions and can think through every ramification of his intents.

The study of physics must deal with all parts and events in all interacting fields. Events resound through fields you cannot see or feel. Yet those events are effectively real. How can you be expected to know anything about invisible events? You look of their shadows. Expand the subtleties of your senses. Believe that life is not separate from the physics of your studies. Turn to your inner resources. Call upon us with your questions.

A boy goes bicycling down the street and turns a corner. His attention is diverted so he accidentally hits a curb. His body is hurdled over the handlebars by momentum from his traveling space tori. He lands painfully onto a sidewalk where his body stops in shock. The finer USUT fields of his body travel onward through the pavement. New USUT Fields replace what he has lost. The interception of bicycle and curb could have been predicted by the neighbor who stood nearby. The intent of the speed and direction of the bike was KNOWN before the collision. It was known, as well, by the finer Universal Field signals that travel out before the action in the form of radii of the boy's sphere, centering within the traveling boy and bike. These radii travel back and to the sides just as any radii of a sphere. At the instant of collision and shock, the radii expand as resisting energy potential is raised. Mechanical rebound is experienced. As the boy is injured a call triggers the instant response of the LAW of RENEWAL so that healing can begin.

Just before the boy's accident he turned a corner. His bike described a section of a circle. The curved section in the ST Field automatically set up radii converging to a center of a sphere which the curve described in part. That center point happened to fall upon a tree. The tree registered an energy gain of specific force. Another example is when a person opens a store on the corner of a busy street, the center point of radii from turning cars falls within his store. The store is constantly energized by these fields, even to a point of frenzy. See **Illustration 4 -5**.

TURNING TRAFFIC
IN TIMES SQUARE, NY

A STORE LOCATED ON A
CORNER WITH TRAFFIC
MAKING TURNS AT THAT
CORNER GAINS EXTRA
ENERGY INSIDE THE STORE

ILLUSTRATION 4-5

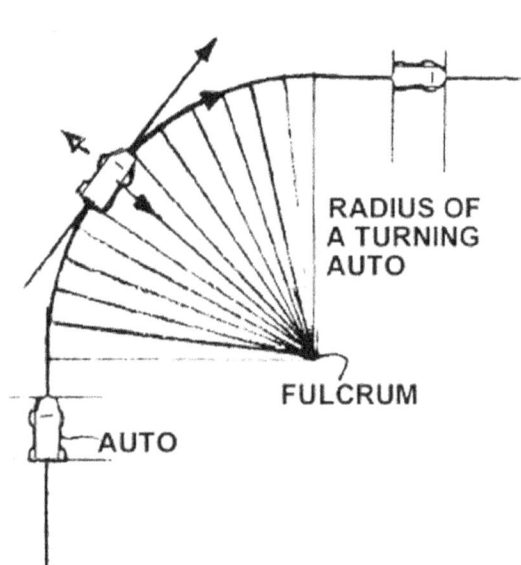

RADIUS OF A TURNING AUTO

FULCRUM

AUTO

WHEN A TRAVELING BODY MAKES A TURN,
ALTHOUGH ITS SPACE TORI OPPOSES THE TURN,
WHILE THE GT FIELD ACTION OF FRICTION
FACILITATES THE TURN,
THE ST FIELD RECORDS THE RADIUS OF THE TURN
TOWARD ITS FULCRUM POINT.

THIS HAPPENS BECAUSE OF PRESSURE OF
TRAVELING TORI PERPENDICULAR TO THE AUTO'S
DIRECTION OF TRAVEL. A HIGH
POTENTIAL FIELD POINT IS CREATED
WHERE THE EXPANDING TORI INTERSECT.

WIND BLOWING THROUGH LEAVES OF PLANTS
CREATES POTENTIAL POINTS IN PLANTS WHICH
STRENGTHENS AND TOUGHENS THE PLANT.
WIND OVER A HUMAN BODY ADDS TO THE
'FREE RADICAL' ELECTRONS
WITHIN THE BODY. ARTHRETIC PEOPLE
ARE TROUBLED BY WIND. AS
THE WORLD TURNS POTENTIAL POINTS ARE
CREATED EVERYWHERE

ONE LANGUAGE

If ALL conditions and events are KNOWN before they occur, then "future" events are also know. A parent is apt to warn a young child of a certain hazard. There are beings around you that you often call **angels** that KNOW more than you do. They may often warn you of hazards, if you listen carefully. The human ability to KNOW events is enhanced as he learns to sense ideas stored up in field centers. Psychic persons seem to get information "out of the air," and in fact, they do. As the radii of your personal sphere expands you can expect to have a wider understanding of your present time. Vast amounts of information can be integrated simultaneously into the human mind. It is not necessary to think in digital sequences. The neighbor who observed the inevitable bicycle accident, feeling helpless to intervene, recognized the boy, whose name was Mike, who wore a red shirt, who pitched the winning baseball game last week, and so on. Much information came together for the neighbor at the same instant of recognition.

When you speak of "inspiration" you may say that the entirety of some knowledge filled your brain at the same instant leaving you surprised and amazed because the idea did not come in words or sentences. All knowledge presented itself at once and that was exciting.

As your sphere grows wider you can focalize more information at once into a vortex in your brain. You will be able to alter the time-space continuum within your sphere. It is possible to construct field conditions to accelerate or decelerate time and potential. This is the energy that a true healer uses. A gifted healer can speed up the natural forces of RENEWAL. A healer can energize parts of a body that are sluggish or have blockages. A healer can convey messages in the Universal language, just as many animals can.

Natives of outback Australia are celebrated for their ability to telepathize between each other over miles of desert. If they switched CB radios their abilities would decline. Telepathy is not digital. It happens in the vortices of the mind. A person who practices keeping an "open mind" can expand his sense of reality and KNOWING. An "open mind" is always accompanied by a large crystal clear sphere. When a Buddhist monk practices meditation, he hopes for a harmonic alliance with a cosmic god source, an understanding without words and the skewed translations that are part of language. He "clears" his mind and his being. A Christian teaching tells of the advantage of being as a little child. Esoteric teaching advises holding no opinions. Not uncommonly an unsophisticated young person is the recipient of inspired religious messages. Most people carry a mental gridwork of biased and scattered information that can cause misinterpretation of Universal language. Then again, most people have based their pride on what they know and what they can do. Since pride is so closely guarded, it becomes almost impossible to replace old information with new. The gridwork is held in place over the vortex like steel bars and the Universal language cannot imprint upon the finer energies of the vortex.

The Universal language has more to offer than the grandest public library and the most informed computer internet. The Universal language is seldom mentioned in a schoolroom or even a Sunday school. The modern world has turned in their psychic strength for very fancy electronics. That makes for a very vulnerable society. Access to the Universal language of MIND is a mandate for any millennial alchemist as he/she may find the self with no place to plug in the toys.

THE REFLECTIVE PRESENT

There are teachers among you who say that everything is an illusion and nothing is real. We will tell you that everything is real including illusion. There is nothing which is not something, and most of it is both synchronous and simultaneous. If you choose not to deal with one reality you will automatically create another.

Any event creates a comprehensive sphere. Each part of an event creates spheres within spheres. Actions and forces within a sphere act as radii, pressure, and harmonic reflections. The overlaying USUT Field sphere will respond with higher octave equality.

Within a sphere there is reflection from the inner side of the walls (surface tension). If the surface tension is built for defense it is like invisible polished steel; everything that is inside reflects back and around, everything that is outside reflects off and does not penetrate. A sphere with softer surface tension can both reflect and interchange with spheres of environmental lines. This is not a metaphor; these spheres are around everything and everyone. You build your sphere as you grow and understand. You can change your sphere as you WILL. As you change and appreciate the LAW of EQUIVALENCIES your sphere will change within seconds.

Why is it humans cannot SEE the spheres and its multiple forces? The human brain is designed like a valve that lets the ST Field cognize in one direction and not the other. To see all the forces acting simultaneously would be very confusing to your brain and impair your ability to make decisions. You mind must learn to focus and to selectively shut out extra events. Fine tuning is part of your current education. You were gifted with sensory selectivity to suit the brain you have at present. That brain had to be trained over thousands of years to integrate the signal input you experience. Your ability to detect vibrations is narrow but expanding. By offering you ideas beyond your present sensory capacity, you will reach for some major biological advances. You can adapt to a bigger reality when you make the decision to do so.

Increasingly now, you will want to have a range of selection of space-time alternative realities. If you were to review events of miraculous healing you would notice that the natural process of healing a body was facilitated and shifted to "fast-forward." The Principle of Renewal was involved and time was encapsulated, all within a sphere that included healer and subject. This is a genuine trick within cosmic LAW. You will want to expand and contract time and its functions.

Your first practice is to focalize the present. While memory is useful and necessary, it can cloud the sense of present. This must be done in full awakened consciousness. Continually clear your old sphere by sweeping it through with an undefined clear sphere of present. This can be done with visualization. Know that you can do this quickly. Die to your history. Bring in RENEWAL to yourself as you bring in the present. Once having found your sphere of the present, practice alternating expansion and contraction of that sphere. That is lesson one.

Why do people like to go on vacations? Vacations are new! They are nonrepetitive. That gives you a sense of present which in turn allows an energy flow of recreation. With imagination you can create a similar vacation experience every day. Some people do that with creative hobbies like writing or painting. Some do it by throwing dice or gambling, or other ways. We suggest that you do it with imaging and moving energy in an expanded octave sphere. As some people might work an algebra problem in their head, you can conceptualize forces and modify their time -space predominances. When you are asleep or dreaming you lose touch with what is present. In fact you are experiencing a present in a very different reality when you sleep. A rested and awake person has an increased sense of present. Full consciousness and environmental integration helps you to identify the nature of present energy. Consciousness is a joyful blessing. Consciousness for the body is activity and exercise. Consciousness for the nervous system is new sensory stimulation. Be careful not to blow your circuits by excessive exposure.

As you spend more and more time in the present and forget who you were yesterday, people may call you a stupid goose. Don't let it worry you. You will have access to a much greater library of information than your memory can provide. THAT WHICH IS KNOWN will become available to you. Beyond the door of the conscious present lies energy in octaves and harmonics new to human experience. Yet, there is never a reason to downgrade the fullness of the experience of human life; it is a marvelous design for a neophyte creator.

Can you see in your imagination the spheres around the trees and rocks? Can you see the spheres around yourself? Flex your spheres, outward and inward. Do you see that each ring intersects with all other rings? Your radii cross again and again with all the others as your concentric rings overlap, just as they overlap in a molecule of gold. Each point of crossing is a living event. Each intersection joins to give you access and sharing to all that is around you. you are ever present everywhere as you fuse your harmonic patterning in merciful alliance to the golden present. Your path widens as your creative adaptation allows your convergence with all. Through the humility of concordance, through the timidity of faith, you may touch the Universal power of Grace. Feel the rhythms and BE in full consciousness.

Your personal sphere may expand, contract and obey the commands of your will. Within any sphere reflection will take place. The signals of our life design will move inside a sphere as sunlight within a cut diamond. The directions of these signals from harmonic pattern as they reflect with equal angles from the inside wall of a delicate surface tension. The harmonic pattern sings your note. This is your key. You create the patterns of your sphere with a geometry of your totality. The patterns carry the theme of your present which can interlock with patterns of your environment. The work of internal patterning is your second lesson. There is much more to say about it as we progress.

Practice modifying the color of your sphere inside and out. Make your sphere iridescent and transparent to match the coloration of the higher, grander octaves of the LAW of LOVE. Design yourself again and again in a new light. As the Universal fulcrum shifts you will want a new balance and not a reinstatement of old credentials.

ST Fields are the genii in the bottle who are always responsive to the will of living beings. It is elemental life. The ST Fields are surprisingly willing to call into play the equilibrating forces of the Universe to substantiate harmonic arrangements. Although the ST Fields obey the LAWS of GOD, and provide hard substance whenever called to so, the fields are not rigid. Their diversity of expression is far greater than all the beautiful flowers that ever bloomed in the gardens of time, and greater than all the sweet features of babies on all continents. There is no end to their dance when the music is played. Fields gather together in a symphony of a living cell and divide and divide on command in miraculous numbers with precise perfection.

CHAPTER 5

VORTEX

AN INNER CIRCLE

A great rose window, set in a gray stone wall of the Chartres Cathedral, casts ruby rays upon the polished floor of the revered twelfth century church. The jeweled light fills an immense religious space from the round window above, over 42 feet in diameter, divided into a twelve part lattice, each stained glass section depicting stories of ancient religious figures. Centered in a round frame is Mary holding the infant Jesus. Queen of Heaven, she reigns over the apostles and priests, kings and soldiers from the Old Testament. Which architect was inspired to create such a glorious attraction? This orifice would be said to magnify the sun, not God, as the center of life. It was designed, without quarrel, as a vortex to the sky.

This sacred geometry at Chartres and other magnificent cathedrals throughout the world has survived wars and weather to remind you that esoteric inspiration is not new to your century. The vortex of the rose window testifies, not to one prophet or queen, but to the cosmic truth of the church of Rome, in an attempt to elevate its purposes beyond the reach of human fingertips. See **Illustration 5-1**.

A vortex describes the equatorial plane of a sphere within its circling circumference which is often identified with the feminine, as it replicates and magnifies, nurtures and sustains. The esoteric view of this vortex of the feminine is that of power, raw power. Axial divisions in that plane describe stabilizing modifications of that power, primarily denoting law. Concentric orbital rings on that plane delegate power to action.

Every religion has an inner circle of keepers of the truths, an out circle of teachers, and an even greater circle of believers in the faith. These circles of people with their delegated power, support the center which is the church. The church is not a facility but an entity of mind. The builders and organizers of Medieval churches were far more interested in spreading the idealistic power of God's church on earth than in the exemplary teachings of the prophet Jesus. Above all, the church elitists

Rose window in
Chartres Cathedral, France

wanted power, both spiritual and temporal. If power was their goal, then a grand vortex would do the job. Ironically, female believers in the faith, in those days, were not allowed to enter the part of the church into which the rose window was built. It can be said that the male religious teachers truly did regard the image of Mary as a symbol of the church itself, virgin, untainted by human feelings, unsoiled by female seductions.

The scintillating colors and grand structure of the rose windows of Europe's cathedrals are, eight centuries later, still a destination of masses of people, townfolks and travelers, who are held spellbound by their history and beauty, who are willing to give up their dollars for the experience.

The story of sacred geometry came into Europe from the Middle East where traditions go back before recorded history. Many groups, among them Masons, tried to bring the sacred sciences from ancient Egypt and Mesopotamia, preserved only in myths and crumbling documents until the dark ages of Europe. The Masons, a guild of builders and merchants, believed that government of the secular needed to be separate from church rule. They gathered what powers they could to fight an underground war with the church that lasted several centuries. In secret places and meetings sacred geometry was taught and records kept. Experiments in alchemy were clandestine as it was used to wage war on a powerful enemy. Gradually, geometry, mathematics and science of chemistry became secularized, just as did the governments of Europe. From the deadly turmoil of misunderstanding, the alchemists and scholars stood up to oppose the shackles of church doctrine. They succeeded through an advanced knowledge of weaponry, plus the general decline of the Church of Rome.

And today scientists are principally motivated economically and politically. Mechanisms of war still take priority. Yet, every scientist knows that truth is his goal and truth is what God is all about. Because of democracy and freedom, secrecy is fading, education is blossoming. Knowledge about cosmic law and power can belong to everyone with an interest and a will to intensively study.

Knowledge about the ST Fields is the essence of sacred geometry. Fields have been misunderstood all through your history because they are invisible. Very few ancient personages knew of them. Yet, the symbols of the fields were carried along with many ancient civilizations by scholars, high priests, leaders, teachers and artists.

What was hidden is now revealed. What was before the prerogative of the few will now belong to the people worldwide. The invisible power of the ST fields and the USUT fields is now presented to you at great risk and promise. The idea that matter is energy and energy is matter was evoked before you with the explosion of the first atomic bomb. Everyone had to know the power of the atom so that freedom and democracy could be maintained. As scientists and technicians are materialistically employed everywhere, the role of alchemist changes. The alchemist is always in the lead; and this time the alchemist must lead civilizations away from a very disastrous materialistic course. The study of vortexes serves that end.

Vortexes will simultaneously replicate all signaled input and enlarge it. Vortexes divide themselves into dynamic, rhythmic wavelengths with unlimited harmonic designs. Every biological creation has use of multiple vortexes. Engineered with skill, a vortex has vast applications and capacitance. Energy gain and creative replication happen in a vortex. It can be the primary tool of every scientist. A modern alchemist, however, needs to know the vortex as a dynamic biological tool within his own being. He must realize, learn to use, the inner circle within himself/herself.

EYES HAVE IT

Looking into the eyes of a lover one sees the life of the world in a small deep pool. It is eyes that surely tell about bright intelligence. You recognize that eyes are a meeting place between people. You will be glad to know that the eyes of the Star people can look closely into your eyes and you will someday look into ours. Beautiful eyes are the vortexes of light and life that can speak without words.

Eyes interpret signals carried by electromagnetic wavelengths of the ST Fields known as LIGHT. They resolve images that the brain can interpret by their design including a variable aperture and a biological avenue for alignments of field radii. Eyes receive subtle vibrations that happen in a field. These coupled vortexes also give off vibration to a field which are not electromagnetic, a field which coexists with the electromagnetic. Your physical structure is replete with invisible organs. They are precise and delicate features far more complex than a molecule, organized in intricate patterns of strands of energy and flowing chemistry. These structures are composed in the spacious ultraviolet spectrums but make chemical connections to the higher octaves.

Many of you have direct experience with the function of these organs during what you call an "out of body experience." One person's experience began with an auto accident when fear of injury caused the "etheric" body to take leave of the mortal body and to hover over the scene, returning only when the body seemed safe. Thereafter, having learned how to disconnect, the "etheric" body left at the slightest threat, like when the auto hit a small bump in the road. At that time the person experienced a second set of eyes lifting upward with full vision then settling in place again once the threat had passed. At that instant the person literally saw with four eyes. These are common experiences that should show you the existence of "etheric" organs that function and are especially connected with vision and environmental relationships. Most people have had clear visual dreams of flight (unassisted) over the tops of trees and houses, simulating memories never experienced in the mortal body with uncanny accuracy. These dreams are recalled from memory generated during "out of body" states. Deathbed experiences and others describe the "out of body" state as euphoric or ecstatic. The "etheric" body is not an emotional body that can be compared to bodily emotions. the "etheric" body is connected by couplings to the mortal body in several places, but primarily near the solar plexus and the top of the spinal column. All this is important to recognize in order to understand that you have the facility to maneuver the vortexes of our own body, and that your invisible body is truly concerned with humanity and the environment of earth. We cannot stress this too strongly. The "etheric" body is not only intelligent but houses the sense of who you are. When you take leave of your mortal body, you will retain your individual knowledge of your soul personage. This experience gives human life true purpose as an educational episode in a being of multiplicity.

Eyes are your windows to the environment. Two eyes focused lets you estimate distances. Two eyes establish a center point with radii in place. Two eyes, more accurately four eyes, give you a "plane of perspective" in a spatial sense and in a mental sense as it teaches you about relativity. You were gifted with eyes because you needed to know your connected ness to the sun and the stars and all life.

Cameras are designed like eyes in terms of their structure but lack the ability to interact with the light source. **Illustration 5-3** shows a photograph of the projection of the sun in eclipse shining through spaces between leaves of a tree onto a floor. Notice that each circle of light is a crescent as was the sun during that time of eclipse. The spaces between the leaves act as apertures to resolve a tiny picture of the whole sun in a short focal length. The aperture is a vortex with orbital transverse waves that can serve to replicate the sun's rays an infinite number of times in correct resolutions. The harmonic images carried in signaled sunshine offer life to most living things because while the images are whole the intensity is reduced. Direct, unfiltered sunlight carries potencies that harm delicate biology. Man developed under the filtering of trees. The delicate organs of the eyes would not have survived a desert.

Your study of eyes will reveal that the eye has multiple functions, such as providing vitamin D and stimulating sexual organs. Understanding four eyes will increase your interest in interconnectedness. Etheric eyes are designed to see in an etheric world. What might they see?

THE GRAIL

More mysterious, more subtly esoteric than the geometry of cathedrals are the legends of the cup of the last supper of Jesus the Christ, which have traveled the world in story and imagery. The Holy Grail, imaged as the cup from which Jesus drank, could be regarded as no more than a memento of an event; except that it is taken as a symbol. A symbol of what? There is a long list of possible sources of its symbology. We shall examine only one of many.

Since the beginning of time, pictorial artists realized they had a double language in relating "field and ground." A mark on a wall meant some people would see a mark rather than the wall. It took a special point of view to recognize that a mark and a wall interact in equal importance and that a wall could "stand for" something with reference to a mark. Like writing with invisible ink, artists learned that they could say secret things by configuring a background to a mark, a "ground." Most ancient art should be looked at to see the background speaking as artists and scholars used codes to escape from political persecution. The cup of the Holy Grail can also be examined in terms of what is in substance and symbol and what is around it in the air.

The bowl of the cup is a half sphere to hold fluid as spiritual nourishment. The stem of the cup narrows to support the idea of a primary axis. The base of the cup flares again to give the cup stability in gravity. The base sits upon a plane of a table. (Many early vessels had round bottoms to rest on sand or a rock.) Esoterically, the table can be understood as part of the cup. The cup with a stem is unique for it describes a double spherical vortex both within and without. Why did the idea of a double vortex become connected with the ideal of Christhood? Why did it survive in legend as an object to be sought after?

A double spherical vortex is a secret reference to the creative power of Christ in the world and beyond the world. See **Illustration 5-2**. The power of Christhood to manifest force is often alluded to in a 45° axis emanating from the juncture of the primary axis and the equatorial plane. This axis is given the title of 'sonship', the effective union of the father and mother principles. You see this invisible axis in the union of four (or six) spheres that surround the cup. That same union is expressed in the Christian cross as a juncture of a primary axis and a secondary axis, but in a more subtle fashion. The person whose hand could encircle the cup (becoming a capacitor) would be gifted with the complete BEING of Jesus, as man and as heavenly son because Jesus had imparted his fullness to the substance of the cup itself. To grasp the cup was to BE the image of son and father in Christ in person. As that event could fall to any worthy man was an astonishing idea, for in olden days it was a law that only the lineage of priesthood could assume the ideal vestiges of Christ. To advocate otherwise was a sentence of death.

A double spherical vortex offers great powers if the whole story were grasped, powers that could be rested upon a civilized plane (the table). Whatever vibrations the cup held in custody would be replicated in any drink it held, giving the drinker all that he needed. Of course, it was a very special person who would have that privilege, just as it would take a special person to draw a sword from a stone. That person would surely have a special FORCE. Or so the story goes.

Single and double vortexes can be seen in religious artifacts from all over the world and in use at any time, ancient or contemporary. There are too many to speak about, from holes in dolman stones to the metal in the prayerful hands of the Tibetan Lamas. That is because a vortex makes manifest unseen signals, unknown realities. They are the rings through which a man or woman can speak to ALL THAT IS.

ILLUSTRATION 5-2
DOUBLE VORTEXES

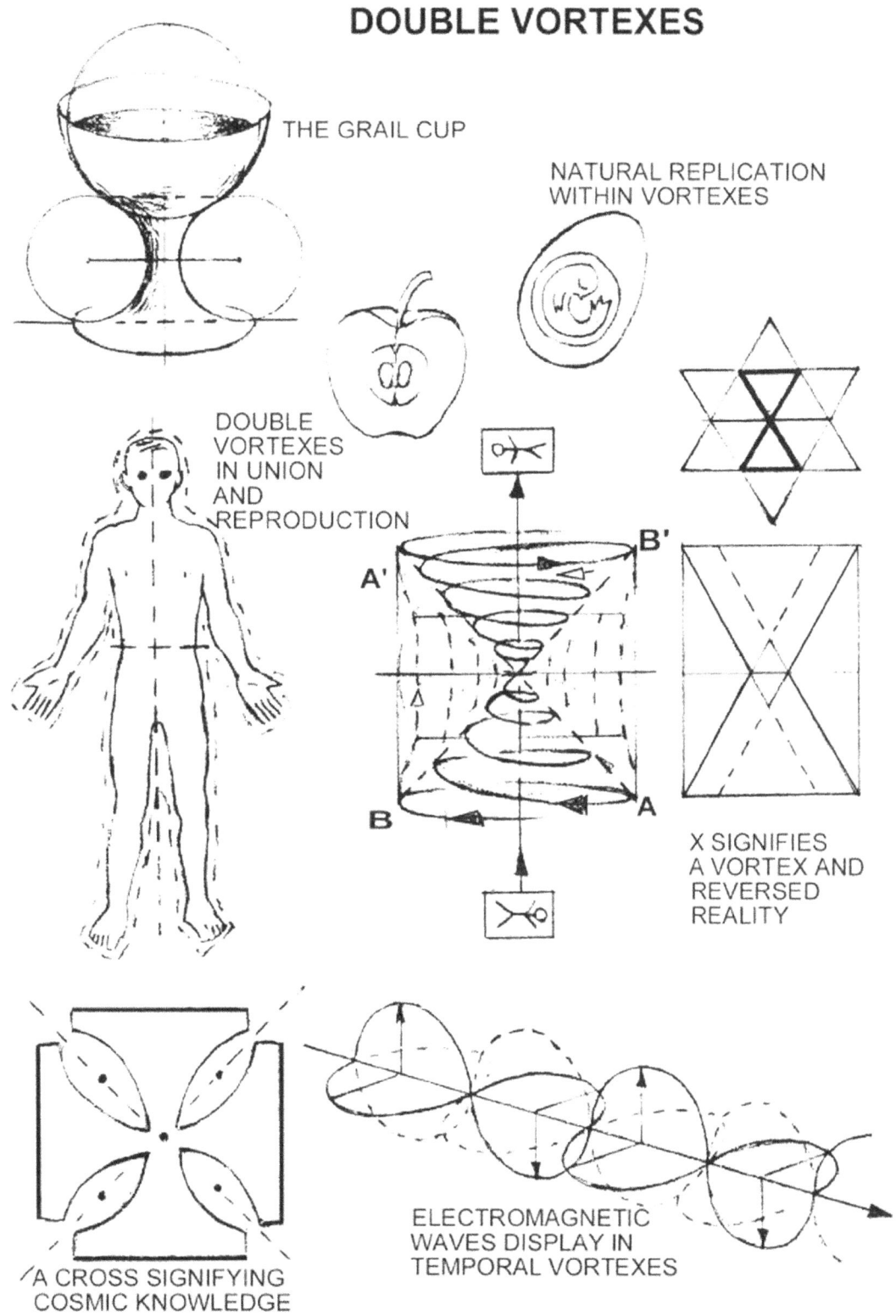

PROJECTION PROJECT

Turning to light, please look at an example of the function of a vortex under the Law of Renewal. Any vortex (which follows the pattern of an electrical toroid) will replicate its greater signal in any number of internal circuits. Here is an easy way to see that happen.

Use a slide projector with a slide in place. The arrangement of a projector's lenses is shown in **Illustration 5-3**. Tape a button to a cardboard with one hole left exposed so that one smooth sided small aperture is open through the cardboard. Project your slide on any screen in a darkened room. (Throw the projected image just a bit out of focus for clarity.) Take up your cardboard with its aperture, then place another white cardboard behind it, within the focal length of the projection so that light travels through the hole. Put your homemade aperture anywhere you like for no matter where you place it in the light, you will get a complete small image on your second white cardboard. The pinhole projection will be complete, accurate and vividly colored. Any place on the focal length carries all the information of the whole.

This exercise will also show you how undertones are always replicates of overtones and visa versa.

How many full images of the slide are replicated within the primary focal length of the slide projection? Any number(n). If one image is electrically removed will another replace it? Of course. How is it that all those undertone images can overlap without mixing up into a jumble? Signals are carried in octave divisions. The measure of the aperture organizes an undertone mathematically to resolve the information of that octave group. Light wavelengths favor certain mathematical groupings. An aperture of about 1/32 of an inch works well. Other apertures, too big or too small, may not give resolution to the organization. Would this happen in a vacuum? Does starlight come to you through airless space and resolved into your telescopes? Once an image is imposed on light by means of a vortex (such as the picture of a slide through a lens) does that image ever change? Not unless the image is taken through a second altering vortex. The lighted image of the slide is two dimensional. Can the same kind of slide projection make a three dimensional image in replication? Your work with holography shows that it can. Holographic imaging is like a virtual image, a show of light and shadow. Can solid objects be formulated this way? Solid structures use the same principles but different mechanisms. Crystals can be easily replicated while living cells are not so easy. Cell division and cloning are now commonly done in a laboratory. Is it not more important to replicate good ideas than material objects. Be cautious about backsliding into material desires. So much of what you desire is an outgrowth of animalism and fear. What do you really want to replicate after all?

ILLUSTRATION 5-3

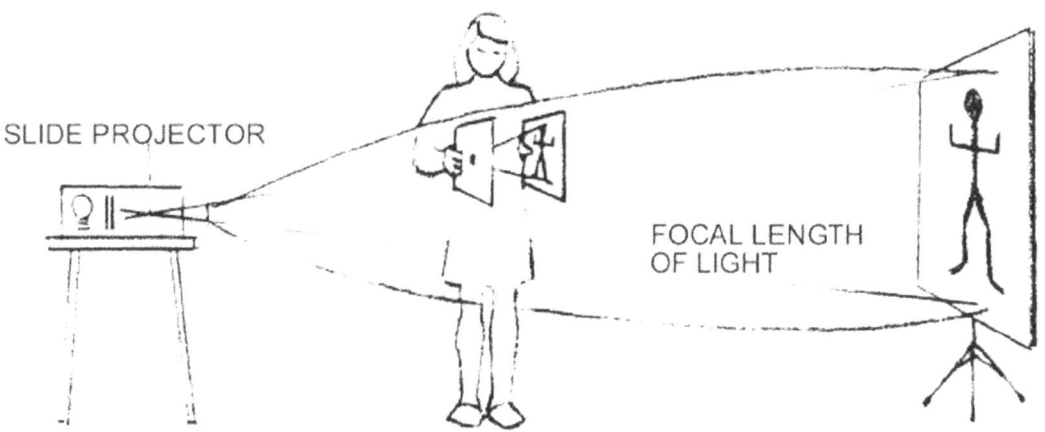

AN IMAGE FROM A SLIDE PROJECTION APPEARS
IN ANY AND ALL PARTS OF THE FOCAL LENGTH DUE
TO THE VORTICAL DESIGN OF THE LENS. RESOLUTION
OF THE SLIDE CONTENT REQUIRES
A SECOND VORTEX (a button hole).

SUN PROJECTED THROUGH THE LEAVES OF A TREE ONTO
A FLOOR DURING A SOLAR ECLIPSE. EACH TINY ORIFACE
CREATES A PICTURE OF ONE WHOLE SUN THAT IS PARTIALLY
BLOCKED BY THE EARTH'S SHADOW

Emeralds are manufactured in a chamber where heat sustains strong kinetic activity in fields supplied with certain gasses. A very small crystal of green beryl is stationed properly in a vortex of the chamber as a seed for molecular patterning. The beryl radiates its exact hexagonal harmonies into very large overtones that fill the chamber and beyond. These high frequency harmonies are characterized by directionally radiating six sided vortexes which divide themselves perfectly into overtones and undertones. Exact applications of heat over exact time periods cause the jewel to accrete, to grow, in orderly replication, to form a boule from which jewels may be cut.

This artificial process of crystallization is like the way emeralds are formed naturally in rock. Chapter 11 will detail the process. Having achieved this engineering feat, emeralds were devalued in the marketplace as only an expert can tell a natural emerald from a man-made one. Diamonds are manufactured in the same way. The expense of making diamonds is so great as to make the process unprofitable, at least at this time.

Jewels are valuable because of their natural properties of energy which shift one frequency to another. However, when men are abused and die from the extraction of gemstones, such as occurred at the diamond mines of Africa, they have only a negative value. When envy and worldly power gather around a simple stone, it is degraded because of negative energy it will store up and be carried by the stone. A green Zirconia can be prettier than an emerald. A high grade clear Zirconia sparkles more than a diamond. Why do we like colored lights? Think about that.

SACRED REPLICATION

In the Bible there are many stores of miracles performed by Jesus. Each one of the miracles was given to illustrate a law of the cosmos. In the book of St. John is told a story wherein Jesus took baskets containing five loaves of bread and two small fishes and multiplied them among five thousand followers at the time of Passover. Jesus then said to them, "This is the work of God, that ye believe on him whom He hath sent." In this act Jesus demonstrated the Law of Renewal. The actual replication of food through the vortex that IS Jesus symbolizes the seasonal work of God to provide grain in the fields. It also symbolizes a magnification of human life to greater resources, a promise of everlasting life. The elevated octaves that Jesus accessed granted him privilege to actualize the law in condensed time sequence. Through the work of Jesus, you were given to know about the higher mansions of reality to which you have been invited.

The same Biblical chapter tells that living beings may signal their image and consciousness in a holographic manner. Jesus appeared as walking on water in a high wind. And also, Jesus appeared at different places, interacting with people simultaneously. You may consider these events as literal truth, as signaling is also *our* usual mode of travel. As you can understand signals in fields and the power of the Law of Love, you can know of many more miraculous ways in which high octave physics can work.

The Bible was not the first book to describe laws of physics. Among the first books were the earliest Hindu scriptures and the Bhagavad-Gita. Scriptures of the Tantra depicted the elemental Symmetric Fields by sexual imagery. Sculptural shrines of the lingam and yoni were placed in many communities as a teaching of life principles. The appeal of the teaching overshadowed the primacy of the signal of MIND as first cause to the general populace. The teaching of Yoga has gained respect for its application of physics to biological self-training and the attainment of enlightenment.

THE SPIDER'S WEB

A vortex is like a toroid. It swirls its currents round and round, not only inward or outward but around. In its center (seen as a cross-section) are patterns of radii and concentric orbital rings generated by the circuit loops around the toroid's circling current. See **Illustration 5-4**.

In the center of a toroid is a pattern that resembles a spider's web. In ancient times, in the Americas, there was a creation story that told about spider woman who created the world with her web. The analogy is surprisingly accurate. The rings of webbing drawn around by a spider are not concentric but convoluted. When currents make a change in the toroid, the equatorial rings move in and out in a convoluted way. When a new galaxy forms its field, arms are outreaching in convolution. So the patterns of a spider's web speak of **creative change**.

The spider, at the center of her web is alert to the insects trapped in the sticky fibers. She captures these insects for food and often hauls them to a cache at center. The web is her purpose and her life. For others it can be an unwanted trap. The spider's web, as a metaphor for the creation of earth, has a second meaning which is threatening. The energy of earth is all encompassing and provides life of a specific design. That energy can ensnare other types of living energy whose forms are vulnerable. The spider produces her offspring by the hundreds at the expense of other species of life. Yet , in the wholeness of things, the spider has her purpose and her glory.

It is the nature of a vortex to seductively draw energy through its center and to impress that energy with its own message. The uncommitted energy around a vortex can be sucked in unless it is wary. Yet, such energy may be looking for a nurturing place and find a vortex to be a home.

The pattern of a labyrinth is similar to a spider's web. To understand the meaning of a labyrinth or a spider web in the cosmic sense we will refer you to Chapter 11. Briefly, materialization occurs when two field spheres of opposite predominances intersect forming a plane of union. The common plane (or chord) is circular. It looks like the union of two bubbles with a plane in between. Around the common plane is a toroid loop which carries a circuit. That toroid generates the pattern upon the plane it encompasses.

If you examine photographs of the gigantic figures drawn on the earth of the Nasca plains of Peru you will see both the labyrinth and the spider placed as identification marks at the end of spacecraft runways. These exacting figures were made to identify work places from the air. The Star People used robot craft in the general shape of a ball or tube, programmed by computers to do specialized jobs. They were able to excavate earth for mining or to moves stones in a path of any size. **Illustration 5-5** are photographs made by a courageous researcher you know as Erich Von Daniken. These are the same tools being used to make crop circles in many countries today. Similar electronic balls are used to get the attention of the peoples of earth by hovering over crowds. They are usually equipped with cameras and other analytical equipment but are unmanned. They are not there to harm people.

Back to the points being addressed; the union between two spheres is an act of creation with dynamic movement toward a full union. When an S Field predominant sphere touches the side of a T Field predominant sphere they intersect at a point that swirls. As the two spheres move closer together the common plane enlarges, drawing the circuit outward in a spiral. The T Field appears as a spiraled line (figure). The S Field appears as the equal space between the curves of the spiraled line (ground). This is the essential pattern for any materialization. They are an extension of what you recognize as the yin-yang sign. They are elementary symbols given to earth's people to help them to understand their cosmic connections. See Chapter 11 for fuller details.

Vortexes are places to actively create or alter signal speeds and wavelengths. The center of a vortex is always slower than its loop. The orbits (or spirals) it creates are graduated in their speeds, slowest at center, more active at the circumference. The speeds are calculable if the speed of the toroid current is known. Within every vortex there is a measure of WILL to counterbalance its effects. WILL proves to be a bridge to the Universal Fields at center through which energy can be cleared. Mother Earth takes care of her own and perpetrates her energy of life. Her imprint has been personified in the Hindu god Siva. She has an important energy to recognize and understand as she is one part of the wholeness. Your avatars have been trying to inform you for thousands of years that you have options. You may choose to leave the nest of Mother Earth and expand your experience at higher speeds. This is accomplished through WILL and an alignment of harmonics that penetrates the surface tension of the outer earth's sphere at a perpendicular. Cyclic timing is imperative and assistance is required. Any person with an emotional superlative condition will be unable to gain the required speed.

ALCHEMICAL MANUAL for this MILLENNIUM

ILLUSTRATION 5-5

THE SPIRAL LABYRINTH SYMBOL IS ONE OF THE OLDEST KNOWN SYMBOLS IN THE WORLD. IT IS THE LITERAL PATTERN OF CREATION ON THE CHORD OF THE SACRED MARRIAGE.

DURING HIS FLIGHTS OVER NASCA, PERU, IT SEEMED TO RESEARCHER ERICH VON DANIKEN THAT ONLY ADVANCED AIRCRAFT COULD HAVE CREATED THE "LANDING STRIPS" ON THE PLAIN. WE CONFIRM HIS OBSERVATIONS.
THE GEOMETRICAL PATTERNS WERE MADE BY COMPUTER CONTROLLED HOVER CRAFT NORMALLY USED FOR MINING.

CREATURE DESIGNS IDENTIFIED CURRENTLY USED AIR STRIPS ON THE NASCA PLAIN. THERE WERE GOOD RELATIONSHIPS BETWEEN THE SPACE TRAVELERS AND THE NATIVE PEOPLES THAT INCLUDED TEACHING.

BE CAUSE

In the fine strands of a peacock's feather you may read the story of creation. Each strand is a series of cells, in groups of gorgeous colors and textures. Strands laying side by side create an exquisite design, each distinctly a part of one whole design. Each design is replicated with precise beauty in an exact number of feathers for the spread of the peacock's tail. And each male peacock is like its father. Each part is separate but intimately part of a whole design in time and space. Patterns of growth are not only directed digitally but wholistically, replicated by invisible spherical vortexes.

It is not easy to talk about spherical, cyclic, and wholistic vortexes. The word dimensionality inadequately describes this principle It will take a leap of imagination to organize the factors of laws in such complex arrangements. It is good practice to exercise your brain with these studies.

Your life and all organs of your body are wholistically patterned by idealogical vortexes. You are part of a pattern of living ideas within the greater law. You are both substance and non substances and both temporary and eternal and so forth. While you are separate, you are forever connected to all humanity, along with its overtoning CAUSE of BEING.

When you think of words like replication, objectification, ideals, family, relativity, you are thinking about vortical causation. These concerns are human. ALL THAT IS is a name indicting relative unity as it includes objectification in BEING. The wholeness of the galactic experience is detailed with the ways in which parts of life relate to the whole of life. You have a solar system within our galactic system which includes the home of the star people. Our small parts to play in the galactic theater are necessary toward the accomplishment of a grand whole idea. You are players on your small stage of ideas. All performances are to be applauded. You are not alone, isolated, or without recognition. Step by step, experience added to experience, we grow into the larger, more complex whole.

Any person interested in Alchemy is dealing with vortical causation. That person is interested in creative pattern imprinting. Once piece of a puzzle cannot be completely understood without the big picture at hand. One actor's line cannot be held to judgment without the whole context of the play. Use caution in any act of good cause.

PERCEPTUAL VORTEXES

Your perception in consciousness is usually ordered in a sequence of digits or dots, each dot having a singular condition. The colored picture on a TV screen is presented by a rapid scanning of dots from the top of your screen to the bottom in horizontal sequence. Your eyes work with your brain in a very similar way.

But there is another kind of perception where signals are simultaneously imprinted within a vortex in your brain and body. You recognize this in times of inspiration or crisis. These images of harmonic simultaneity are directly comparable to responses within a field sphere. They act under the Law of Renewal and Energy Gain. The brain acts to summarize and analyze patterns of digital input because of this ability. It is the centralizing feature of the brain where multitudinal, multi-dimensional aspects of information are overlapped for assessment of interrelationships. Coding occurs as in a moiré patterning. Any such centralization happens within a vortex of specific energy design, within a half-sphere with an equatorial axis around a primary linear axis. That divided half-sphere locates specific harmonic numbers of smaller numbers. The area is much smaller than the brain itself but the patterns are replicated throughout the brain.

Simultaneous vortexes within the brain allow a person to abstract, deduce, reduce, and relate all knowledge within the consciousness. As you know, brain function is critically reliant upon blood flow carrying oxygen and other crucial chemicals to the cells of the brain. Within the oxygen is the active ST Fields which are directly utilized in the brain activity. Once released from oxygen by caloric burning, the ST Fields quickly pass out into the environment. New Oxygen must be immediately supplied to provide the necessary fields for life. The Yoga call these fields prana and it is delivered through breath. ST Fields can also be delivered to the head through etheric vortexes.

The more complex the brain function, the greater is the need for fresh supplies of ST Fields in the circulatory systems. Bigger brains circulate more blood and often have more intelligence. But ST Fields move freely through the skull. Yoga practices demonstrate field exchange without breath. Most humans will do best by improving their breathing techniques.

Remember that the ST Fields around you are there for you to use as an energy resource. Those fields are extremely responsive to the calls of your expectations, your desires, and the totality of your needs. As you take responsibility for the purposes of your life so will follow the strength of your self will, the ability to command yourself.

A child learns to walk by way of the desire to get from one side of the room to the other. A child is not aware of every muscle and bone in its legs. A child wants! It has a "user friendly" brain that does the rest on command. You will activate your invisible body by wanting. Wanting "stuff" is a low level motive. Wanting ideas, knowledge and experience will get things moving. After wanting, the details of "how" will begin to fall into place.

You can want to be healthy. You can want to eat candy all day long. The two programs conflict. Your planning needs to have carefully considered priorities. Your wants may conflict with other people; how shall you prioritize that within the overtone of harmlessness? You may want to change the world. To change the world, a powerful, willful focus must be activated to change yourself. To do all that you will need the help of compatriots of the higher octaves.

The need to hold power over others is an immature expression of a weakness to command the self. When self will is fully realized, others will be moved by inspiration and example. **To steel from others the development of self-will is to defeat the evolutionary progression intended by MIND.**

The need to create a world of mechanical robots is a show of evolutionary failure. A mechanistic society is extremely vulnerable. If advanced technology is the single aim of any society or persons they are marching backwards. When the glamour of industrialization wears thin, groups of mature people will emerge to demonstrate a fuller capacity for humanity. Such personages are seeds that dwell among you today. Look for them.

As you cup your hands together in prayer, thereby constructing an S Field capacitor, think about what you ultimately want. And think BIG. Think big plans that include everyone into a better world. What is it you want?

MEDITATION

There is a quiet peaceful place in the whole of you. It is a still, balanced, harmless place. You often notice it in spring when the blossoms are pink and little birds fly in and out of bushes. It is a place of appreciation where the growth of interconnectedness can occur. Sitting in a blank place with a blank mind will not encourage a sense of presence or wisdom.

You did not come into this world for the purpose of leaving it. You preplanned your path to integrate with this blue planet, this time and space. You are moving faster now, miles and miles everywhere, everyday. That is not accidental or a result of greedy motivations. The fields are shifting; your speed will help you to adjust to galactic change. You are thinking faster now, including more factors into daily mysteries. The global economy challenges your language and skills. It is all necessary. But then, in contrast, there must be quiet times to hear the whispers of your heart and the vibrations of the cosmic melodies. Take a drive in the country, slow to let a startled rabbit cross the road. Sit under an aged tree whose roots tangle with a billion year old rock and listen. Listen for a new language with no words, no spellings, and none of the usual meanings. Be an open door to fresh moments when the Law of Love is singing. Things have already changed, time now, present, life in sudden splashes, in whirlpools that take your eyes by surprise. Listen as the wind speaks because this life is here for you, just as you planned it. Breathe deeply to your roots.

LESSONS FOR CHANGE

Look at your abdomen. There are circles of life there. There are powerful lower vortices. In most of you the solar plexus sits comfortably swaying to your daily exuberances in a very lazy way, not doing what it was designed to do. Of all the chakras in your body, it is the most powerful center of spherical life. The solar plexus broadcasts your life spheres outward in concentric rings to great distances. It activates all the other chakras of your body and energizes your organs.

Exercise your solar plexus with diaphragmatic breathing. With each in breath, expanding your chest, press your sphere outward as far as your imagination allows. Include with your expanding rings a signaled pattern of a six pointed star. This will begin your practice of harmonically imprinting your inner sphere and outer sphere. Send your star out over the mountain tops as you expand the fields in your lungs. With each exhalation tighten the muscles of your solar plexus. As your tighten all your abdominal muscles, put a vortical swirl of energy around your middle, rotating to your right so that energy is driven upwards through your spine, from the ground upwards and out the top of your head. Work and relax your body in periodic harmonic moments.

A vortex with circling power draws energy through it (directed as in the galactic right hand rule). The speed and direction of the circling power determines the speed and direction of the fields drawn through the vortex. This is your facility and your privilege. You can create a vortex when and where you choose in your body, carrying signals imprinted by your will. Self healing can be accomplished in this way. You can create a palette of sufficient power upon which to receive the signaled imprint of your universal compatriots. Or you can empower each chakra in your body with great energy in the same way.

Exercising the vortices of your own being is to be essentially your own practice. It will be your own experience driven by your own will. We can assure you it will add energy to your life in ways that will surprise you. No apparatus is necessary. No purchases. No credit. Only self direction and self discovery. Group work serves the advanced student only, because self determination is the key to the mastery of life energies.

Add to your vortex exercises an overdwelling of harmlessness. There is no reason for defense or offense. No attack at the human level can reach you. Your progress toward overcoming the fear of death is to know that your personhood cannot die (although your might die to flesh and bone). Life continues.

When you infuse your sphere with the attitude of defense, a hard shell is created on several layers of your sphere. Your surface tension acts like steel. Beyond it you cannot reach or see. All internal signals reflect back to the self. All external signals bounce off the surface. External communication can cease. Imaginary, remembered beings take up positions on the shell and become active ghosts of fear. Too many people actually live in this condition.

As you move your sphere outward and inward in breath, you will need to keep these spherical rings delicate and transparent. The rings need to be clean and beautiful, full of radiance, love and joy. The radiant rings can gently move through everything, trees, buildings, soldiers, mountains, etc. Nothing can come up against your transparent sphere unless you use it to deliberately build a wall. Adding fear to the sphere slows the currents. Universal love is the nonresistant course that establishes equilibrium, that prepares the sphere to engage with high octave energies.

Vortexes in your head face all directions. Your head serves as a broadcasting station for signals and a receiving station for signals. There are biological mechanisms to open and close the energy vortexes in your brain as well as to circulate energy around them. These centers are responsive to the command of your will. The vortex called the "third eye," centered just above your eyes, is familiar to you. Revolving it to your left will draw field inward, to your right field moves outward. You can command and designate it to revolve at very high speeds. Each speed will alter your energy state and receiving capacity. Full consciousness must drive the commands.

If your spheres are transparent your sensitive radii can carry finely tuned harmonic experiences through to your consciousness. (You may pick up an ancient coin in your fingers and experience a virtual story of the days in which the coin was passed as tender.) The condition and colors of your spheres will precisely determine your interpretation of the reality around you. Altering those conditions, you may perceive a drama as tragic or comic. Vortexes are "eyes" designed by MIND to perfect a recognition of truth.

WHO ARE YOU?

The expression of WILL takes the form of a toroid. It imprints a priority upon all other ideas and activities. It is ongoing, sometimes outlasting any valid usefulness. WILL sets a pattern in motion. Its design and speed, impressed with the heat of passion, engenders power and directional force. Every person spends a lifetime learning how to manage WILL, then is sometimes disappointed when he/she feels that their personal WILL is overshadowed by another, or by accident or fate. The sense of personhood and personality growth comes form a flexible but stable will. The power of BEING WHO you are comes from the confidence of believing that your will is effective in everyday life. Varieties of mental disruptions arise when a person's will is deformed for whatever reason. Personal WILL imposes balanced or imbalanced perspectives upon a person's experiences. In addition, legal or societal WILL adds to the equation of personal adjustment. Some people find it practical to change their name when their will is realigned to a new goal, because a person is so closely identified with their name and their WHO.

Quan Yin is replicated as a vortex of mercy in an ancient drawing. Both male and female, he/she represents the multiplied grace of the Universal Fields.

The equilibrium of the equation of personal consciousness will benefit from an attitude of harmlessness and an identification with cosmic Law of Renewal. Your WHO has the power of command to stimulate your bodily centeredness, axial circulation, and cosmic alignments.

When a person can say I WILL WANT, let him/her believe that the reference is cosmically aligned to a galactic definition of GOOD as it applies to the essential spiritual self. The vortex of power is TO BE CAUSE in a wholistic coordinate system. Take a position of responsibility shared with the axis of MIND. Power comes <u>through</u> the self. If power were to come <u>from</u> the self there would be exhaustion. Personal power and the physics of cosmic law are never separate, they are both center and sphere.

Here we can direct you to a portal of reference. Art and mathematics hold to a secure cosmic axis like a corn stalk holds onto the earth in a high wind. Equations are relatedness which reflect back to a certain, a true, causal beginning. From that place of interlocking, new ideas are integrated, new designs spring outward. Before you begin to argue that the old ways, old axioms, are withered and dead, let us point out to you that you never knew the old laws. No one told you. No one told your families or your churches or your scientists. These days will reveal the biggest leap that civilization has ever taken because what you are learning now will change everything you do. The shadows are fading, the veils are being rent. You will hear the unspoken words of truths that have always been.

If you can walk into the mysterious heart of a cosmic vortex, a toroid in full expression, you can know as MIND knows. In your dream place, stand on the plane of the sun's equator and observe the energy of axes that move through their rounds of circulation. Allow galactic voices to guide you through the invisible lights that swirl across illuminated numbers in degrees. Observe and carefully step into the concentric rings of divided energies. Align your will with the strength of your secret governor so that clouds do not sweep you into dark shadows. Wait, only if you are not sure of WHO you are beyond the edges of consciousness.

CHAPTER 6

WINDS OF QUETZALCÓATL

FORCE IS A FEATHER

You now have the essential knowledge of work and force. Step by step you can unravel the puzzle of magnetism, electricity, and force. As you work in the fields you may have a faint memory of the ancient god figure of Quetzalcóatl, the plumed serpent. As the soft breezes blow through your shutters, sit quietly and meditate, imaging colorful magnetic circles. Remember stories told by the art of weavers, winding one colored thread over another to reveal the creation of the world. Understand how beautiful colored bird feathers became a symbol of the links between earth and sky.

Vagaries aside, open your mind and attention to the interweaving tapestries of the fields. It is necessary to begin with tedious reviews so that mental processes can bring clarity to the diagrams of the activity of forces.

The S Field is equal and opposite to the T Field and visa versa. The two fields must always stand side by side in equilibrium. Any action in one field will be met with equal and opposite action in the other. The forces of CLOSURE and expansion are your primary forces.

Space tori form around both S and T Field predominant particles. Space tori also form around fields carrying signals (which will be discussed in the chapter on light). Space tori evoke the mechanical forces that provide power to your industries. Tori also move continents, build mountains, and blow the rain clouds over your crops. A space torus forms around every motion, every force, and form in accordance with the "right hand rule." The black arrows, indicating T Fields are used for clarity, but wherever there is a T Field flow, there will always be an S Field flow in the opposite direction indicated by an open-dashed arrow. An N direction of a magnet's flow indicates a T Field flow in that direction. A circle with a dot in it means a flow coming toward you out of the page, whereas a circle with an x in it, or a single x, means a flow going away from you into the page.

An iron atom finds energy of motion within its nucleus. The center of any atom is of a high octave and responds to a signal from a high octave energy source. The signal in an iron atom causes it to develop a space tori with appropriate rotation and charged particles. The rotation and alignment of these iron atoms in an iron bar act like an engine that drives the surrounding fields through the fine channels of the iron crystalline lattice. The T Field is driven out the N end of the magnet, the S Field is driven out the S end of the magnet.

Illustration 6-1 shows a diagram and circulation of a space tori. Such a diagram is helpful in mapping out the position of directional forces. Always remember that a tori will show a primary circulation in the opposite direction of the motion that stimulated it.

Illustration 6-2. The flow of field in the magnet causes space tori to form along the length of a magnet and around each field line flowing in its looped flux pattern. The presence of the space tori guides the loops from one end of the magnet to enter the opposite end. Historically, this action is illustrated by a serpent taking its own tail in its mouth. Ancient images from Teotihuacan in Central America depicts the serpent with feathers surrounding its neck. It is called the plumed serpent who manages the wind. Other serpents are shown with two heads facing each other or two snakes intertwining. Feathers illustrate the presence of the surrounding, circulating space tori that translate orbital current into force.

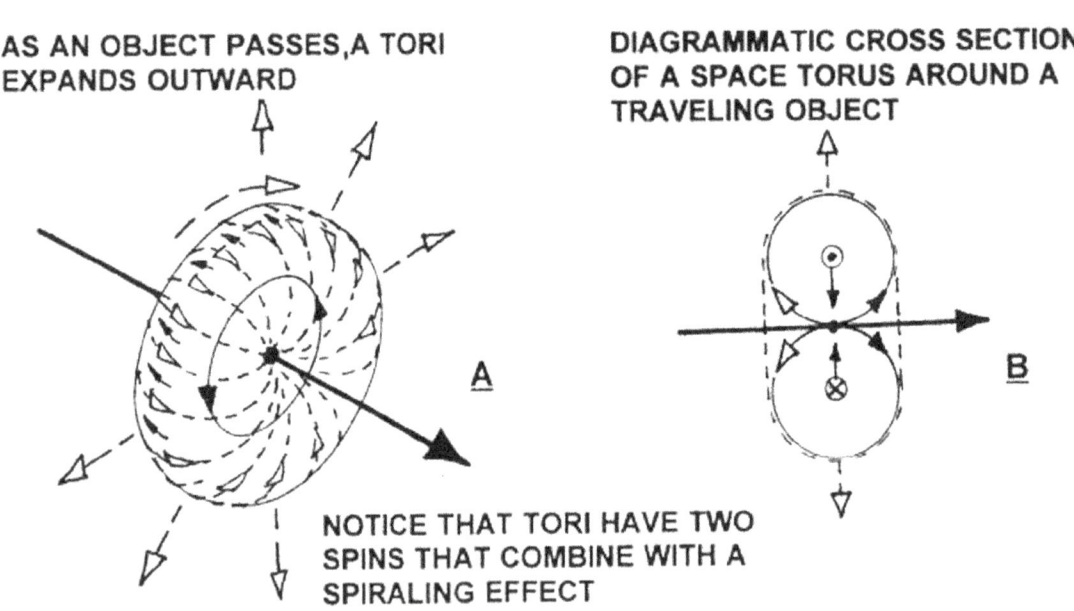

ILLUSTRATION 6-1

Each compatible space torus in and around the magnet joins together in overtones. Central currents within the overtones describe center and orbits. All orbits are fixed in place by pressures within the tori. At the center of each tori a particle will form equivalent to an electron or proton. If the orbit changes as a result of a change in position of a field flux, the particle is released to seek its own path of conduction. Refer back to **Illustration 2-3**, then compare what is said concerning circuit loops to what you now know about space tori.

Illustration 6-3. The circuits inside a space torus establishes an equatorial plane. A primary circuit resonates to become a series of concentric circuits, alternating S Field and T Field predominance. We will call this orbital, equatorial plane our "magnetic dinner plate." The chapter on light will tell you about its colorful attributes. As current in a wire increases its voltage, the size of the torus increases. The diameter of the "magnetic dinner plate" increases. A constant flow of current in a wire shows a steady presence of rings around that wire. When current flow ceases, the rings disappear. But, even without a power source, there is always a very low level of current flow in a wire that has some path to ground. That low level of electron flow comes from electrons that form and discharge along gravitational lines.

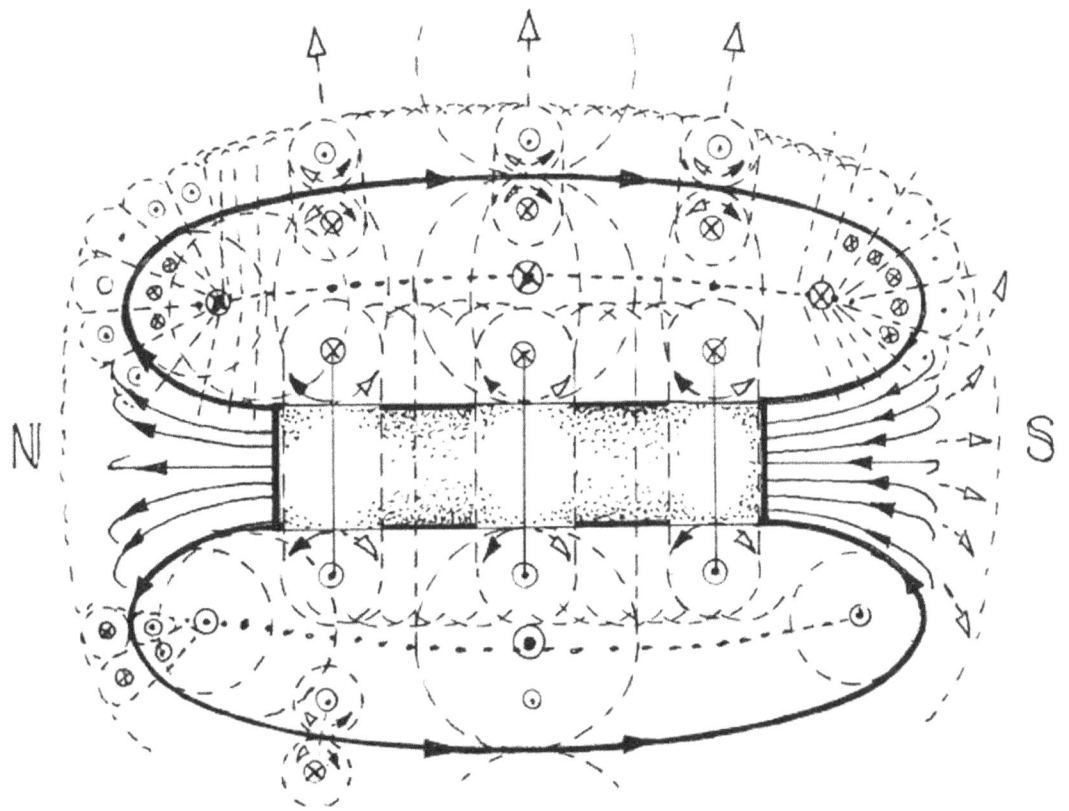

A MAGNET SHOWING FLUX FIELD WITH OVERTONING TORI

ILLUSTRATION 6-2

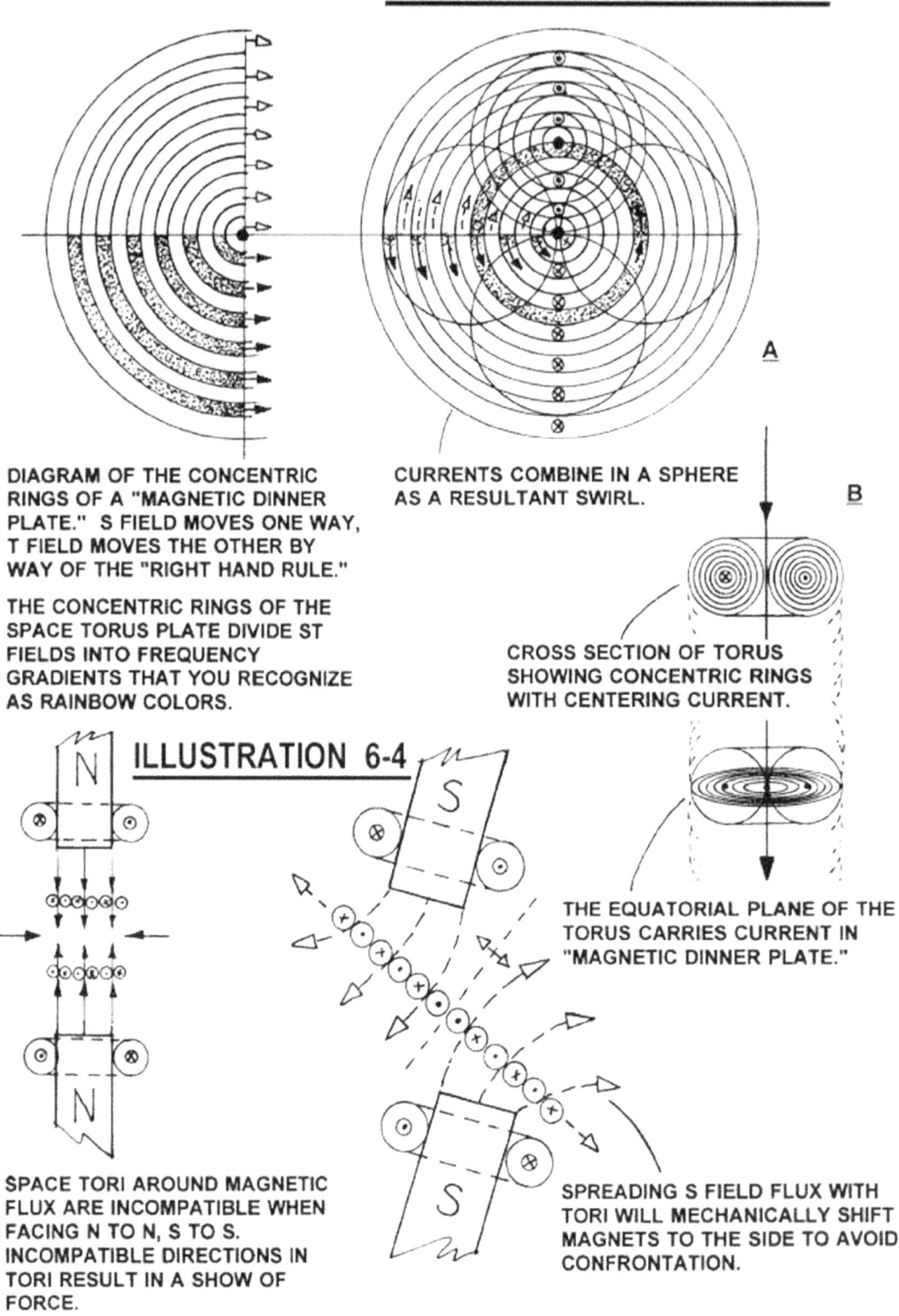

Illustration 6-4. When mankind noticed that one magnet pulled or pushed against another, they suspected that they had a clue to a mechanical world. When they found that magnetism and electricity were linked, the world changed. The new science empowered the world. Electric motors and generators came into being in the blink of an eye. Particles and magnetic flux had a strange resultant called force of motion or "electromotive force."

The natural pressure of space tori are the persistent result of traveling fields. The pressures of an S Field torus are always outward, while integrating with a characteristic spin. S Field pushes; T Field pulls. When two S Field ends of a magnet are placed together, two pushing forces meet head on. In addition, there are space tori around those fields that have incompatible spins. The pressures in the tori cause a shift, a motion of avoidance, with sufficient force to move one or both of the magnets. A force was evident that man could use to help him work. Small magnets made very big news! But, until now, mankind has not realized that most of that force comes from directional pressures in space tori that manifest the necessary space-time ratios for "electromotive force."

Illustration 6-5. Your textbooks show you an expression of force as it is in effect perpendicular, not only to a flow of current, but also to a magnetic flow. When a metal rod is placed between the north and south poles of a strong magnet, the rod is thrust up and out of that field flow. Pressures in the space tori in the magnet's flow acts counter to the space tori around the metal rod. Incompatible tori yield forces that push the metal rod away.

Space tori are around every magnetic flow line and around every group of lines in overtone. In a place of flux between the N end of a magnet and the S end of a magnet there is a bulge in the flow because S Fields push outward. When a T Field particle is introduced into this flux, it spins to the periphery of the bulge in a spin OPPOSITE to the direction you would expect from the "right hand rule." It can be shown that small particles choose to flow in the S Field orbits of tori along their directional paths. That indicates that S Field predominant circuits have a speed and harmonic compatibility with very small particles.

Illustration 6-6. Overtoning is a constant occurrence with space tori. Overtoning increases mass and CLOSURE, as two or more tori come together, acting as one, and forming a new center point. Experiments show that two current carrying wires, with currents traveling in the same direction create magnetic rings (magnetic dinner plates) that pull together in overtone. The wires attract one another. The rings are, of course, space tori that come together in overtone. Any vortex has the cohesion of overtoning in every part.

It is infinite undertoning that allows a whole to replicate itself any number of times in a vortex. Overtoning and undertoning do take place in any compatible direction around tori. When lines of force and spins are incompatible and opposing between tori, internal forces of opposition push them apart.

ILLUSTRATION 6-5

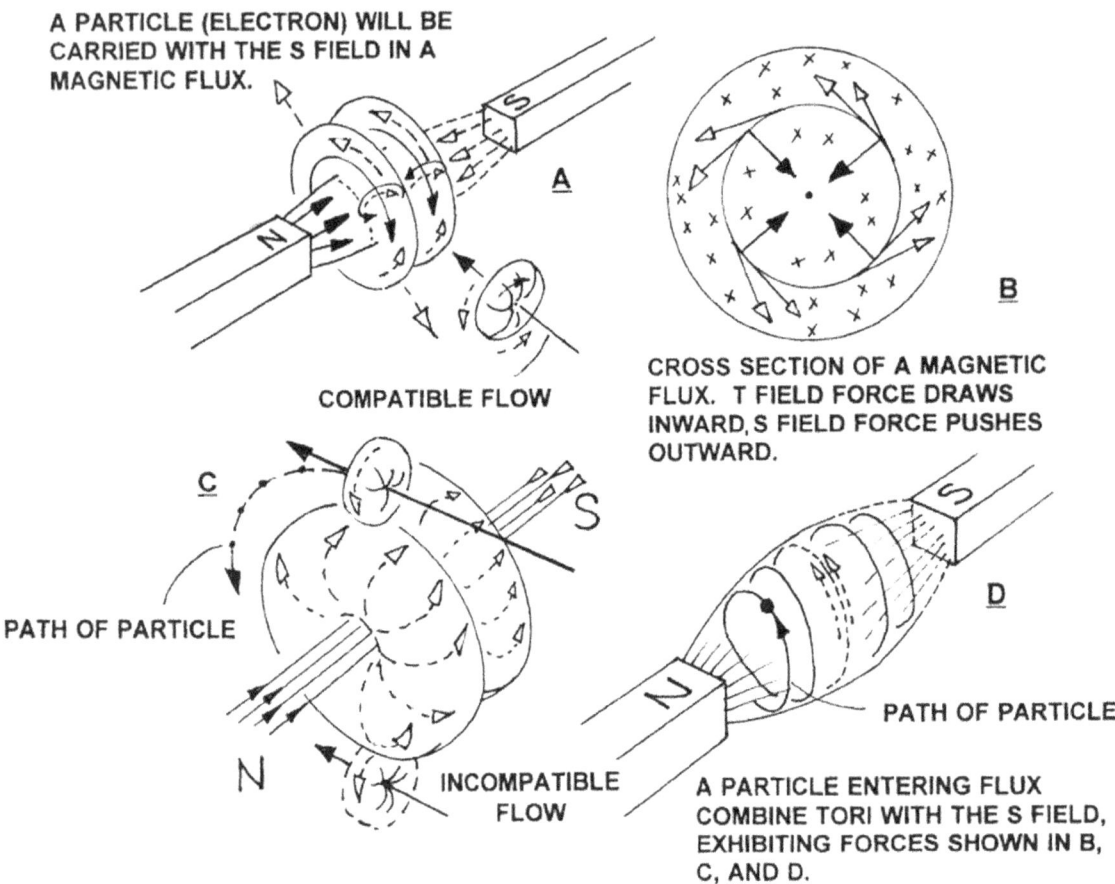

INDUCTANCE

Illustration 6-7. Inductance was described by Michael Faraday in simple experiments, easy to reproduce in a classroom, yet essential to understanding the power of electric currents and magnets. Faraday wound a current carrying wire next to a dormant wire around a wooden dowel to measure what would happen when a switch was closed on the circuit hooked to a battery. **Illustration 6-7A,B,** shows how the experiment produced an opposite moving current in the previously dormant wire. The same happened in wires wrapped on an iron ring as shown in **Illustration 6-7C,D**.

Heinrich Lenz, working in the nineteenth century, perfected experiments combining magnetic field motion and electric induced currents. He succeeded in showing mankind how to induce a current in a wire by moving a magnet through a wire loop. His work with small magnets led to the

ILLUSTRATION 6-6

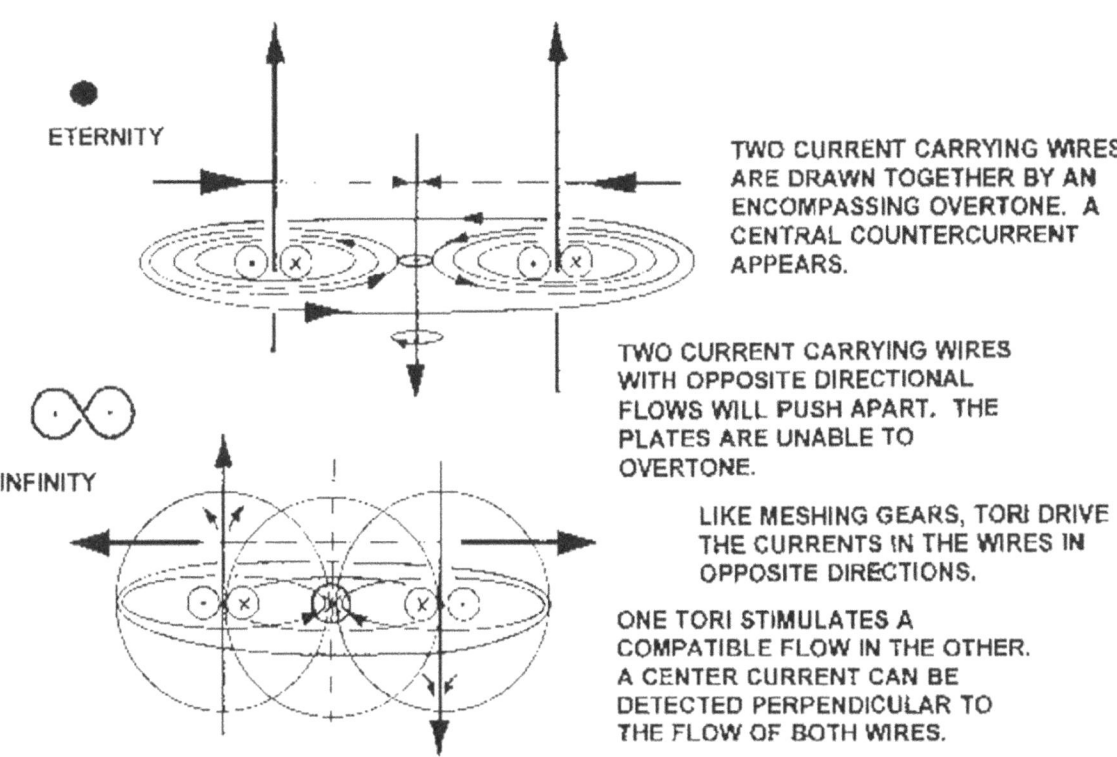

design of giant generators, such as you may see installed on the great dams on lakes around the world. The falling motion of water turns the giant magnets which then produce currents in conducting cables. A simple natural phenomena lights up the cities of today with electricity. The electro-magnetic relationship is the work horse that powered the industrial revolution and then went on to power the electronic communication systems that make your civilizations flourish.

> **Heinrich Lenz states his observations as follows: The direction of an induced current is such that its own magnetic field opposes the original change in magnetic flux that induced the current.**

It worked! A magnetic field induced a current in a wire. But surprisingly, the direction of current flow did not uphold the "right hand rule." Lenz did not know why or how it happened, but no matter, he had current when and where it was needed. We will now show you why this is true.

ILLUSTRATION 6-7

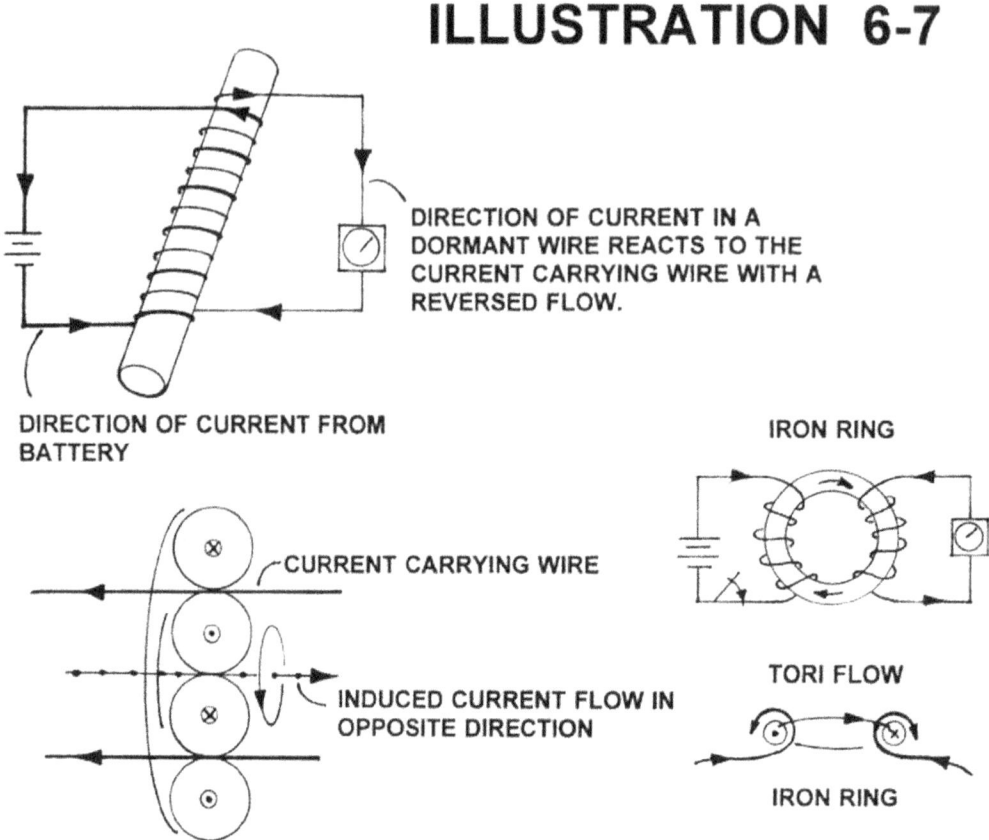

Magnetic action is dependent upon the vibrational speed in equal and opposite flows of the S and T Fields. Current is defined by the motion (speed) of particles known as electrons or protons, as they are taken away from the centers of flux circuit loops. Magnetic loops create particles, but those particles are not set free unless the magnet moves, or unless the circuit loops change position. Whenever there is movement space tori are present. It could be expressed that space tori capture the "freed" particles and take them away on their own circuits. The presence of space tori around a moving magnet will induce tori in a nearby conducting wire at a speed like the initiating tori, often in an unexpected direction. The conducting wire picks up the free particles and carries them in circuit, but the conduction direction is that of the S Field tori corresponding to the south end of the magnet.

Illustration 6-8 A,B,C,D shows the relationship of magnetic flux to current discovered by Henrich Lenz. The diagrams indicate the direction of movement of the magnet in relationship to the wire loop. Lenz, in his day, was not aware of the presence of the S Field itself, with its directional flow.

Illustration 6-9 A,B,C,D shows the S and T Field action of the magnetic flux and also the S and T Field predominant space tori that envelop both flux and the magnets' movements. We bring to

ILLUSTRATION 6-8

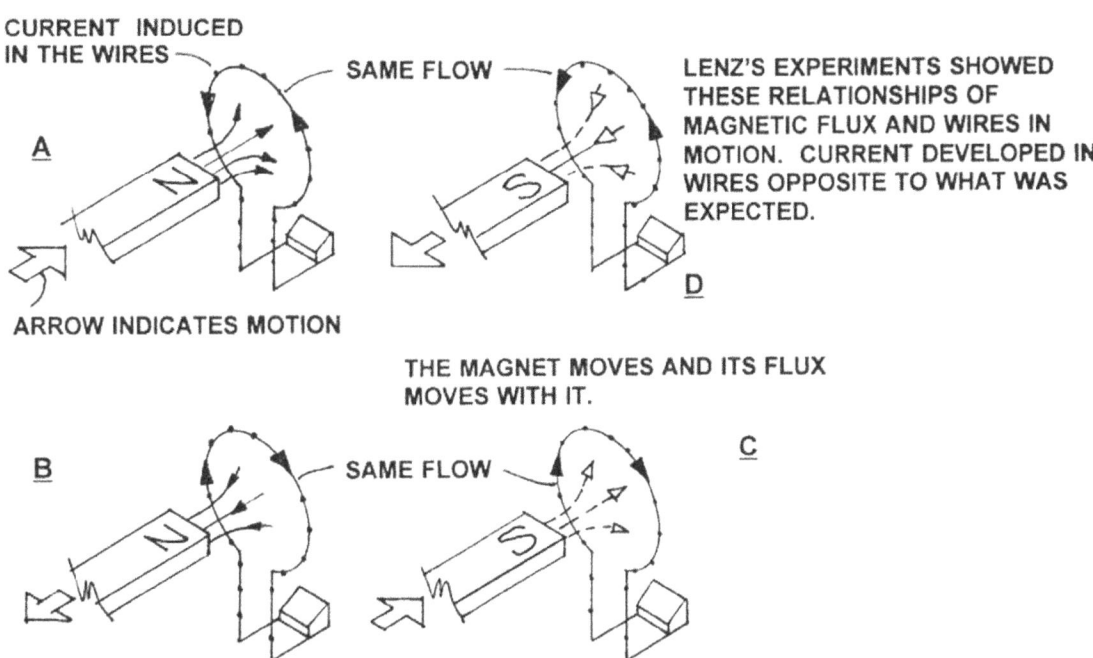

your attention that the space tori we have been describing up until now have been the tori around traveling materials of T Field predominance. These tori spin responding to T Field components of the materials and they revolve in the direction of motion. They follow the "right hand rule."

Refer now to **Illustration 6-9C**. As you might expect, a space tori also travels around a south end magnetic flux, just as one does around a north end magnetic flux. The south end tori around a south end of a magnet will show an S Field predominance. It, too, revolves with the S Field flow and follows the "right hand rule." Now, you will see that the S Field torus and the T Field torus appear to be the same considering their directions, but they are not the same. It is important to recognize the difference in these twin tori. One carries a stronger T Field charge. They are easily distinguished by your radio antennae.

When the motion of a magnet opposes the direction of its flow of flux, the flux follows the motion. Its tori shows preference to the motion of the magnet itself.

The diagram in **Illustration 6-10** begins with the idea of the north end of a magnet (A) approaching a wire loop (B). The magnet is in motion and therefore in the process of creating electrons in the orbit of its tori. When the magnet approaches the wire loop the electrons within the

ILLUSTRATION 6-9

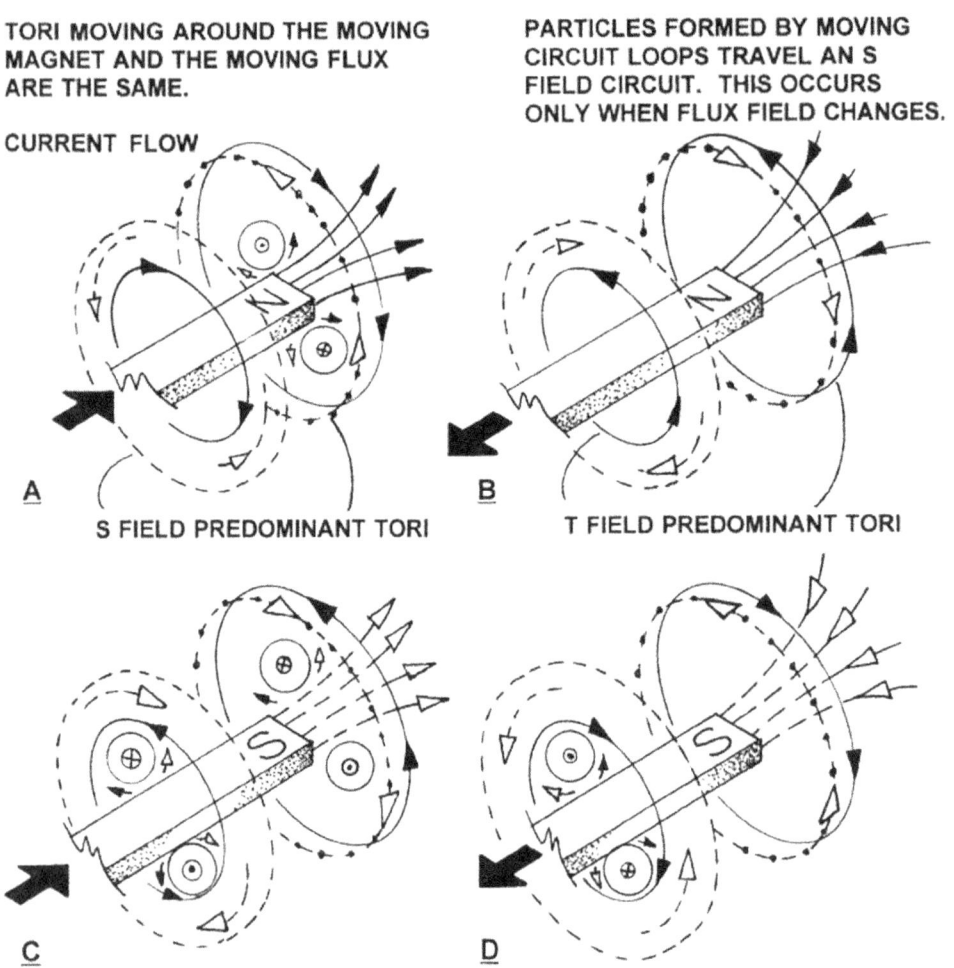

tori jump onto that wire and travel with the current flow. While the magnet seems to be approaching the wire, the wire is also approaching the magnet. Although you may not see the wire move, it is moving with regard to the motion of the field flux. There is an invisible plane (C) where the two motions in CLOSURE are apt to meet. That is the plane of equalized tension where two forces meet one another in a stand-still. Because the wire loop (B) is polarized with regard to the field in which it happens to be, all parts of the wire take a north-south direction (no matter what the metal may be). The torus of each tiny part overtones together making one large torus the envelopes the entire loop, (just as it does on the magnet). That torus circulates just as it would if the south end of a magnet were to be approaching the loop from the opposite side. Because of that, the direction of the current in the loop is as shown in (D). The force of motion of this torus is directed as shown at (E), opposing the force of motion of the magnet (A). This is the phenomena noticed by Heinrich

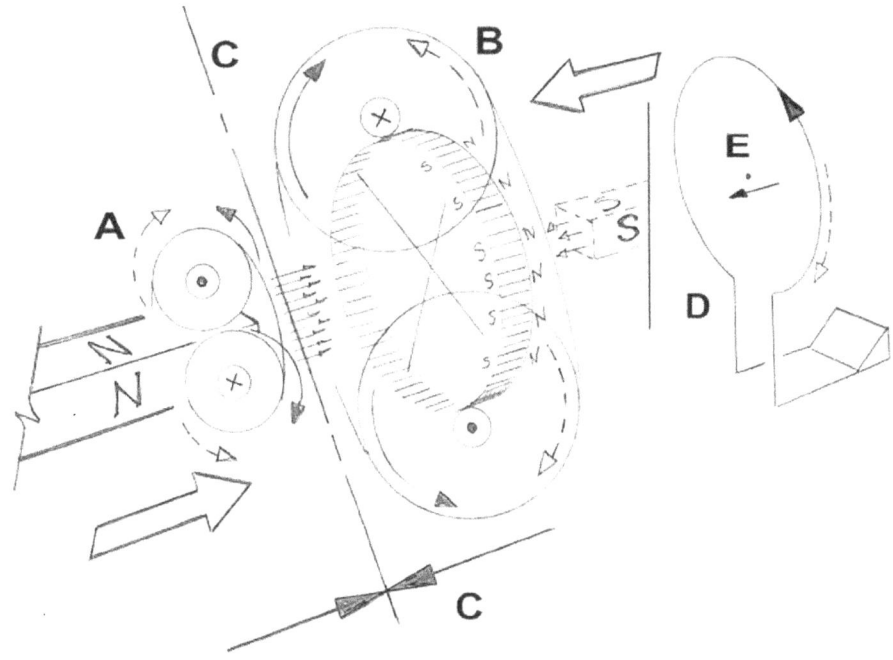

Lentz but for which he had no substantial explanation. The events were chalked up to the idea that nature always tries to conserve energy, and that in fact, no elemental energy is ever gained or lost.

Nature certainly wishes to conserve energy. Nature also wishes to make more energy at every octave level. Through agreements with the Universal fields, the ST Fields manufactures energy of its own kind.

A DEMONSTRATION OF PRINCIPLE

It is easy to understand that the flux field of a magnet intercepts the metal loop adjacent to a magnet and that a flow of current can be detected in that loop when the magnet is in motion. But what is the cause of that current? Why is there such a phenomena as an induced current at all?

A wire, moving in relation to a magnet and its flowing fields, acts as a whole in terms of its overtoned tori which surrounds the entire loop. As shown in Illustration 6-10, the wire finds itself to be in the same place as the center of its motion torus. It is placed in the orbital juncture of loops of the torus, exactly the place where a motion torus manufactures electrons. As motion spins the torus more electrons are produced and then cast off to be picked up by the grounded wire. The faster the motion is repeated the more electrons enter the wire.

The orbital path centering a torus is both S Field and T Field, side by side, moving in opposite directions. The S Field is producing an equivalent to an electron. Each is spiraling at a specific speed and potential determined by the combination the strength and quality of the field and the motion within the field. This experiment is a clear demonstration of electrons being formed at the confluent center of such a motion tori, which then spiral off in a closed circuit. The laws governing the reasons for this have already been stated. This unique situation that produces these useful particles called electrons has all to do with field speed and vibrational harmony.

Any wire, of any material, will develop a current when placed in a field flow. You do not have the instruments to detect all of them. A quartz crystal used instead of a magnet can also produce a current, but not a current of the quanta called electrons. Leaves on a tree are designed to flutter in a breeze within a gravitational field. Doing their dance, they develop motion tori energy which is then delivered to the whole tree.

As the archer shoots an arrow at the target 100 yards away, an equal and opposite force jumps away from the target, passing the arrow from the opposite direction, and stabbing the archer at the same time the arrow strikes the target. It is an S Field predominant force and compatible with the human being. The archer is exhilarated by the strike, having gained new energy. How does the target know an arrow is coming? It knows by way of signals which travel faster than the speed of light. It knows the intent of the archer who has chosen a target.

A BIT ABOUT WEATHER

Since the earth is continually in motion, space tori are always present around everything. New space tori are formed when speed or kinetic energy is added to an object. Space tori are a resultant, not a cause (usually), but their pressures can and do alter speeds and modify their causatives. One space tori can overlay another without changing the first, that is, unless the two space tori are in the same speed octaves with similar vibrations. Two similar tori will add or subtract from the other if they are in the same phase. Space tori are slow to shut down when their cause has ceased. Their energy feeds back into the causative action to prolong it, showing the condition you call momentum.

Because S Fields rise up from the earth, buoyant particles are made airborne. Within the earth, motions of heat, circulation and CLOSURE cause currents in the GS Fields, both inside the mantle and up through the crust of earth's skin into the atmosphere. Earth's rotation causes GS Field to circulate east to west. And, the energy of the sun constantly is added to the kinetic movement of S Fields on and in the earth. When S Field moves it makes weather.

Space tori are strongly effected by signals of hot and cold. Heat will cause tori to spin faster, thus imparting motion to the traveling object causing the tori. Cold will cause tori to slow down, to freeze, to diminish its diameter while increasing its pressure. Warm space tori around air and ocean currents help to distribute heat into balance around the globe.

Illustration 6-11 shows the relationship between the rotating earth and its resulting tori. During the summer in the northern hemisphere, the sun penetrates the earth with signals and currents that cause air, land and sea to heat up. The sun's rays directly hit the earth above the earth's equator because of the tipped axis of rotation. Heat, added to the powerful space tori that band the equator, increases kinetic activity as summer progresses. At the equator the earth rotates at 1,041 miles per hour and moves west to east. Because the earth is widest at the equator and traveling fastest, its space tori are stronger there and diminish in a gradient toward the north and south poles. There is a strong torus around the earth's axes, both north and south. These tori are noticeable by the ringed position of aurora borealis.

ILLUSTRATION 6-11

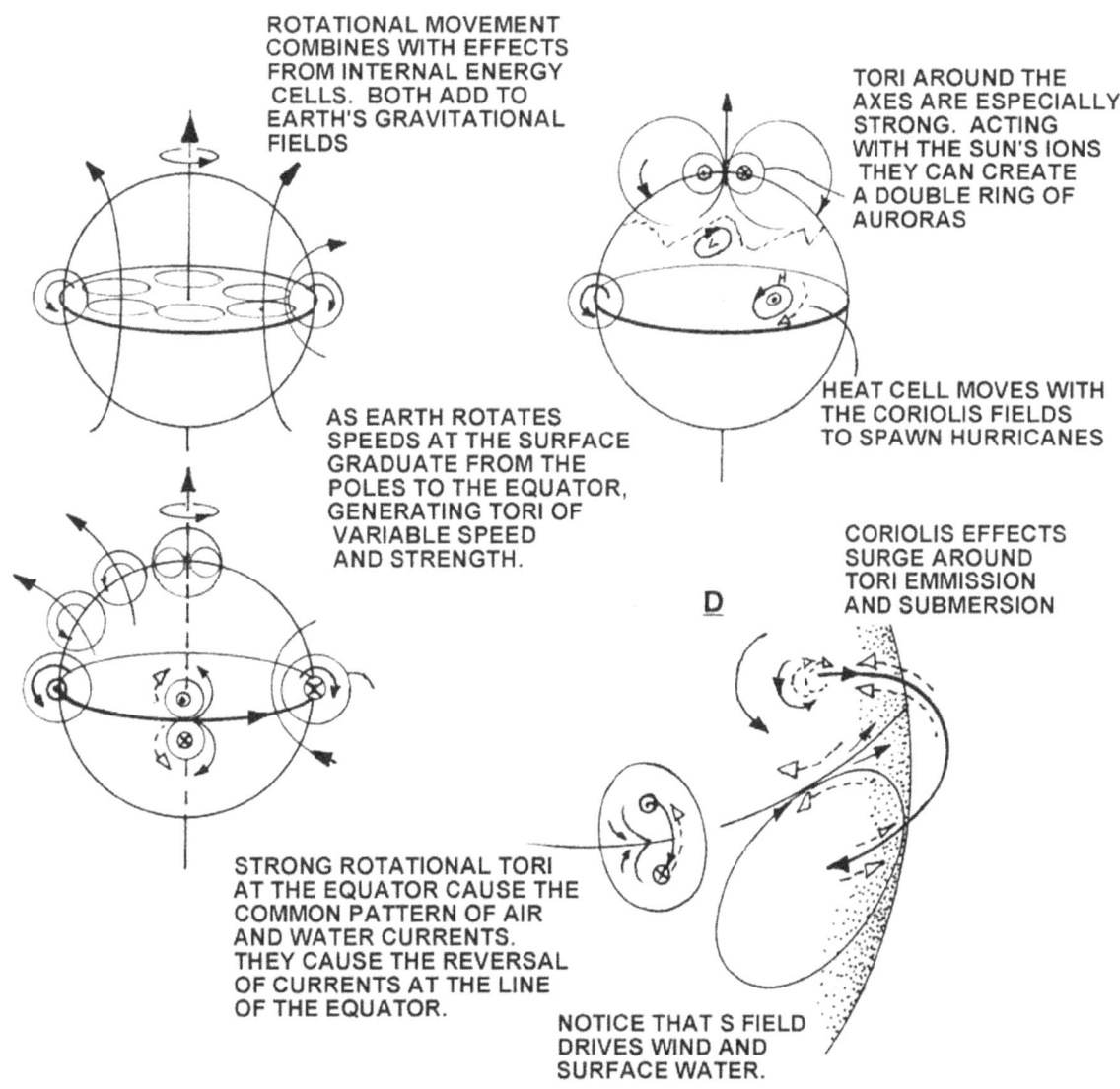

Space tori embrace the seemingly calm currents of the equatorial oceans. To the north and south of the equatorial band, the winds and waters pick up speed. And turbulent counter currents and winds, east to west, brought travelers in small craft from Africa to the Americas on the "easterlies."

The electrical current centering the equatorial space torus moves north to south, plunging into the earth to the south and out of the earth to the north. This counters the force of the GT Field lines above the equator, but matches them to the south. Circulation around this torus current produces a strong effect you call the "Coriolis force". S Field above the equator swirl counterclockwise and S Fields below the equator swirl clockwise. T Fields affect water, S fields move air. **Illustration 6-12 A,B** shows the general circulations of air and water over the planet. Heat and space tori are the driving force of these circulations.

Hurricanes develop in the Atlantic Ocean as heat builds up in the waters off Africa at the end of a long hot summer. Heat and space tori combine to increase the swirling of massive areas of water and air. Cells, or bubbles of field pressure, form with axes for heat transfer from ocean to air, from equator to polar regions.

Illustration 6-11A shows one common location on earth where hurricanes begin during August. **Illustration 6-12A** shows a simplistic map of common ocean currents. Just above the narrow band of currents traveling west to east hurricanes can spawn and be carried on currents moving east to west. **Illustration 6-12C**, shows the top view of a hurricane wind pattern and **6-12D** shows in cross sectional diagram the directions of wind and wave flows from a typical hurricane cell. Space tori, accelerated by heat, strengthen their axes. Around the primary axis space tori are tightly wound. Around the upper axis space tori ferociously winds clockwise, drawing air downward like a spout and condensing moisture from air under high pressure. At the ocean's level the air is pushed outward, thick with water and wind. The water below swells under the pressures of the S Field spins and moves waves outward in surges. The power of the storm is floated with the overtoning cell along with the water currents up the North American coastline into cooler waters where the intensity of motion in the cell is diminished. The cell is selfperpetuating as long as heat persists and land masses do not interfere. Hurricanes will show you much about the ability of space tori to sustain momentum.

A storm is said to have a "low" barometric pressure. Clear weather is usually associated with a "high" barometric pressure. Air pressure and field pressure are being measured. The "weight" of air is not a factor in the readings. Barometric gauges read similarly to an altimeter, lower (more) at sea level, higher (less) as altitude progresses. There is a pressure gradient in the GS Field from sea level on upward into the atmosphere. That increasing pressure in the GS Field is because of its response to the densely packed lines of the GT Field nearing the center of the earth. A barometer is the best instrument you have to measure S Field pressures at this time.

ILLUSTRATION 6-12

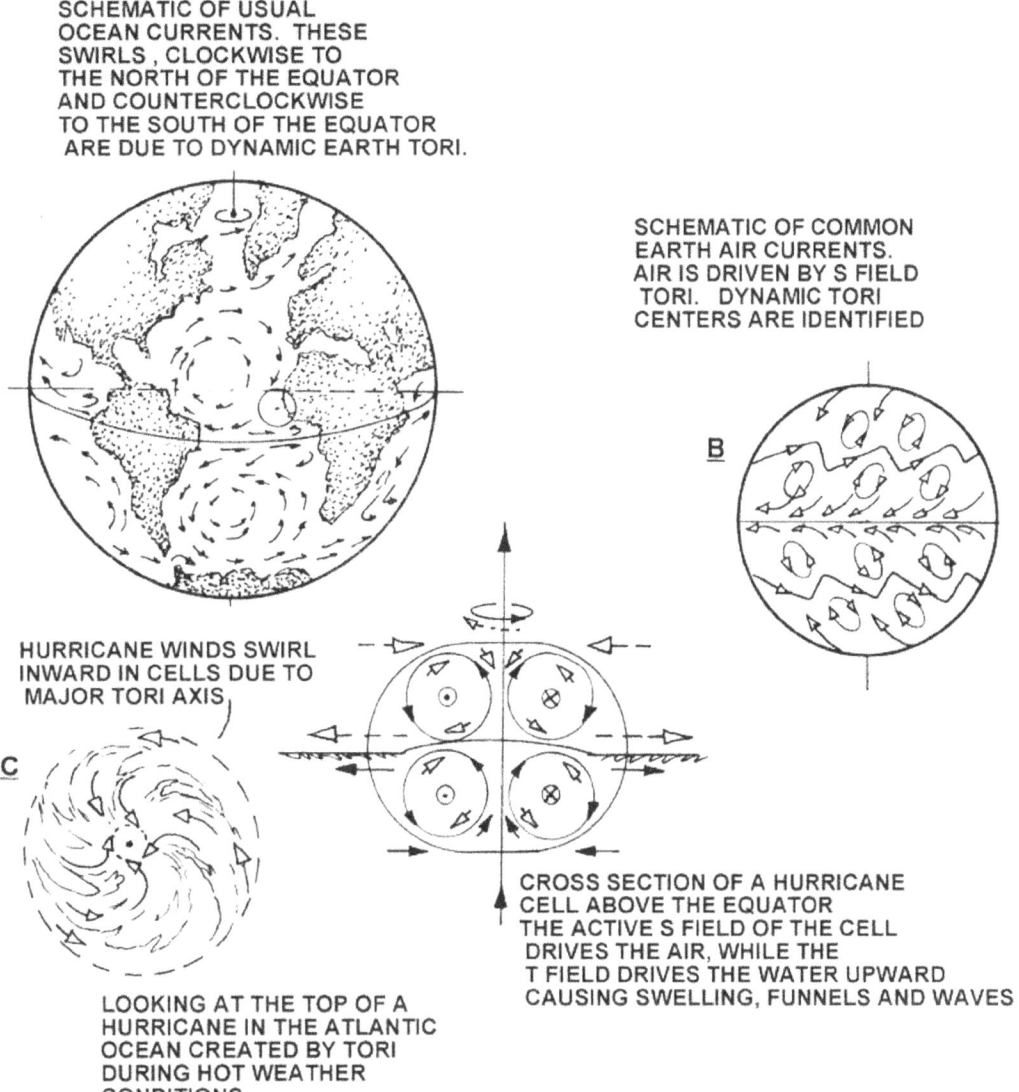

As an aside, here is a joke about a truck driver who often passed through an inspection station. Each time he approached the weighing station he would stop his truck, run around to the side and bang on the wall of the truck. Then he would proceed to drive through the scales with a happy face. Later, another truck driver, having seen him do this a number of times, asked his friend what he was up to. The first driver explained that he often drove an overloaded cargo of canary birds from Mexico to Albuquerque. By banging on the side of the truck all the canaries became airborne. Therefore, the weight did not record on the truck scales at the inspection station. You can think about that clever truck driver and ask yourself how you can expect to weigh a column of air a mile up in the sky. If a gaseous molecule is floating, what is it floating on? How can you join all the air molecules together to read out a collective weight? Why of course, they are joined by GT Field lines. But then you have to subtract the factors of buoyancy to find a weight. Can you stand under a cloud floating in a blue sky on a sunny day and get a different reading than if you stood under the blue sky? After all, the cloud has water vapor in it which conceivably is heavier. The barometer measures S Field density and is a good way to estimate weather conditions. Pressure is not the same as weight. Variations in GS Field pressures can give you a clue to how the earth works.

Illustration 6-13 shows a photo of a very familiar stormy sky. Wind motion and pressure changes make for a dramatic symphony of tori, undertones and overtones, swirling and dancing in the sky above. That magnificent motion tori action produces electrical charges which congregate in damp air. Exchanges in ST Field predominances and charges will crack as lightning. Dynamic circuits suddenly move the air in rolls of long wavefronts. When an equilibrium is finally met a peaceful sky returns. If you could see S Field you could see those tori. The picture could be overwhelming if it were not understood.

Space tori tend to move materials in a straight line. It is a wonder that things do not go flying off the surface of the earth. They often do in tornados. The more powerful GST Fields step in and impose their downward pull sufficiently to pin you down to the surface.

Collective overtones of space tori may play a part in making a planet take a spherical shape. In the formative stages of planets you have a gathering mass of materials coming together along radial CLOSURES. The earth and its solar system progress along its galactic orbit at approximately one quarter million miles per hour. The planetary axis begins a rotation in response to its parent sun. Add the progression of the earth in its solar orbit at 66,500 mph and you can see that you are traveling in the cosmos at a very fast pace, all the time with appropriate space tori. While tori line up along an axis, they also overtone into a sphere with the assistance of CLOSURE at the center of the axis. As massive materials begin to gather at a center place, the tori increase, stabilizing the spin, interfunctioning with the secondary equatorial axis in harmonic dimensions.

Centrifugal force is clearly seen when a plastic body is spun on an axis. Since the pressures within a torus are equalized, they mechanically maneuver substance to travel in a straight line. The substance in a spin exhibits this force outward at a tangent to the axial spin. The stronger force of CLOSURE acts to prevent this spreading. Molten materials and gases exhibit the planar action of centrifugal force better than solid objects in a spin. Dense materials are resistant to the outward action caused by tori. Your centrifuge apparatus was designed on these principles. As the earth formed, it built up heavy materials at its center with a gradient of lighter materials toward the outside of the sphere. The heavy iron-like center accepted its assignment as the center of gravitational force, the center of cosmic breath.

ILLUSTRATION 6-13

THE INVISIBLE TORI AROUND HOT AND COLD FIELDS
STIR UP SEASONAL STORMS THAT CAN VIOLENTLY
INTERACT WITH EARTH'S TORI. THE TURBULENCE
CREATES ELECTRICAL CHARGES IN THE AIR.

The secondary axis of a sphere is a primal force. It is inseparable from a primary axis. Its character is to spin as it is essentially an S Field axial plane. You must make a distinction between the secondary axis of a sphere and any space tori that is likely to occur on that spin. In the same way you must distinguish between an orbital plane and a plane exhibiting space tori. These forces all work together but act in individual ways. Space tori are the direct instantaneous result of motion in either the S Field or the T Field.

Space tori can show a T Field predominance or an S Field predominance. In a case where you may have a partially isolated S Field flow, like at one end of a magnet, or in some of our propulsion vehicles where the S Field is used to maneuver around planets that are without atmosphere, the S Field flow will have a slightly T Field predominant tori. These tori have particularly strong electrical rings. They can be a factor in engineering. While tori are the result of some field action, they are the cause of other actions, many of which we have mentioned. Tori interfere with a union of CLOSURE between a traveling object and gravitational forces.

ONCE UPON A TIME, MANY, MANY YEARS AGO ---

The photo in **Illustration 6-14** is of an architectural detail of the Temple of Quetzalcóatl in Teotihuacan, Mexico. The wind god, Quetzalcóatl is represented by the plumed serpent. Carefully carved feathers at the serpents head curl backwards to represent the invisible space tori. The snake represents the powerful electrical properties of the ST Fields. Often, two snakes are shown twining together in regular undulation. Other carvings may show two snakes with faces butted or heads in equal but opposite directions. But, the serpent with sharp white teeth of Quetzalcóatl tells about a particular, singular field that makes the winds blow. This is not a primitive message. It is a way to teach a practically illiterate people about a very sophisticated cosmic principle. Before the building of the final temple at Teotihuacan, the cosmic principles were known, but kept secret except from a few scholars and priests who had heard of them from their fathers. Although the priesthood could read and write, they guarded their most profound secrets by the oral traditions. Through war and bloodshed, the traditions were lost, the knowledge was lost.

The other mask shown in the sculptured frieze is that of Tlaloc, the rain god, who stabilized the four quadrants of creation. Through his anthropomorphic representation is seen the corpuscular structure of the ST Fields. The little squares are not corn, but field corpuscles arranged true to their usual pattern. Tlaloc is shown with four eyes and two ears in spherical shapes with holes at the centers. They are the electronic vortexes that carry the messages of light and sound and understanding. Water is honored as it falls from the sky from Tlaloc, and from Quetzalcóatl. The honoring of these gods, these principles, is recognized as prayer by the star people who have not forgotten the earnest homage of the Mixtec and Toltec people and their forefathers, and those steadfast persons who have kept the traditions as best they could throughout the centuries of deprivation and suffering.

The vitality of the Space Fields and space tori returns to those of you who can recognize the real lives behind elemental principles. You are connected with those lives, those spirits of benevolence, every hour of every day. The winds that blow over your homes and fields, and brush past your sun kissed cheeks are speaking to you of what is old and what is new and now. The winds tell you about the oxygen filled seas of atmosphere in which you live. As fish lives in ocean waters, you live in seas of S Fields with oxygen and air. Winds tell you about flight and changing times, and about how to adapt to the expanded octaves of the space fields that will release you from the bondage of pain.

Because of fresh oxygen produced by plants and made airborne in the GS Fields, you are privileged to breathe air and the ST fields with it. In breath you find life. As air passes into your lungs the flow gathers tori. Space tori propel the heat that drives your bodily engines with full power. Space fields fan the flames of the calorie burning of food and minerals in your body. Just as a flow of oxygen feeds a fire, space tori blow upon the coals of your bloodstream.

In ancient times warriors learned to run like the wind over great distances and rough terrain. Running was their prayer and their means to strength. Their lungs enlarged for more intake to give them endurance. It was a good idea and a good lesson to follow. You may have heard of the "high" gained by a long distance runner. Notice that the design of your lungs requires flow, not just oxygen storage. Flow induces tori. The swirling tori transfer the oxygen into your bloodstream.

Incidentally, the peoples of ancient South America knew all about wheels and elected not to use them. The philosophy of their leaders taught that mechanical devices crippled personal development. Wheels were forbidden to the poor. These were not ignorant, underdeveloped people. However, in those times the societies were decisively stratified. Democracy had not yet become a social option. Education was a limited privilege. Unfortunately, as has happened in many societies of olden times, when war caused the demise of the leadership, the details of knowledge were lost to the populace.

How was it that the ancients of Peru and Mexico came to have such an accurate knowledge of science and astronomy? You have already guessed. The origins of those people intertwined with visitors and settlers from other stellar systems before your recorded history begins. We, the Star People, have always been teachers on your planet. We have always been instrumental in your social growth. Growth now requires a worldwide shift to democratic social structure, a shift that can stimulate the development of personal freewill. Humanity grows as a whole just as a child grows to maturity. Collective freewill is equally as important as personal freewill. When the personal and collective freewill can express GOOD as a planetary-galactic whole then the human society will have matured.

THE RIGHT PLACE AND THE RIGHT TIME

When Albert Einstein tried to navigate his theories through a variety of "frames of reference," he was trying to work through the forests of octave zones associated with speeds. What happens in one octave may show little or no effect in another. Any interfacing octaves effect one another as a result of common harmonies. One octave can never obliterate another. Effective mechanical laws will only be operational in a limited group of octaves. Your mechanical studies have, therefore, been enigmatic.

One space tori will not replace another. This is true because space tori are a result before they are a cause. One space tori can promote the power to move objects, produce momentum and centrifugal force. To understand space tori you must trace back to an originating source. That source is invariably an idea from a creative MIND. FIRST CAUSE is the single player who is the point fulcrum of all co-existing laws.

For instance, radio signals traveling from distant stars are not obliterated by light signals crossing paths. Radio signals can be altered by other radio signals only if their frequencies match. Because this is true, you have a wide choice of radio and TV stations to tune into. A set of ST Field corpuscles can accurately respond to thousands of signals simultaneously. The radio program and the fact of its being in the field can be traced back to a responsible human being who thought about it.

Space tori come and go quickly but they are also part of the signal carrying systems. Transiting signals create tori, and expand the coordinates of wave action. Their staying power sustains the experience of time. Space tori join one minute to the next, noting the signal, the power and glory of each.

MECHANICAL FORCE MEANS SLOW MOTION

Return your thoughts to Chapter 2, where magnetism and electricity are discussed. We have been introducing ideas about symmetric field physics one at a time to give your mind a way to move from one field activity to another. In Chapter 2 we had not yet talked about the effects of tori around a magnet. We must now confess that it is the specialized space tori due to motion that helps a magnet to demonstrate attraction and repulsion. Tori also release electrons that can be lured away to follow a wire circuit.

Tori around a field flow swirl at high speeds, matching the field flow. Tori around a magnet in motion travel at slow speeds **Mechanical force results from fields moving at slower speeds**. Electrical flow is the result of fields moving at higher speeds. When combined, tori can do more than one job.

Tori around a normal flux field flow has the capacity to move another magnet through CLOSURE. That motion is accelerated when tori due to motion are added. That total motion, in turn, can drive a steel axle. The axle can drive the wheels of a golf cart. An ST Field equation is stepped down for power in series, slowed down to provide the needed force. Tori around a shifting flux field flow (as when a magnet is moved) affects electron particles by slowing them, enhancing their particulate unity, and by moving them into safe passage within an S Field torus to a place of stable lodging.

Mechanical force is a result of ST Field flow moving at slow speeds. How slow? What are the electromechanical characteristics of force?

Motive force is assigned in the electromagnetic spectrum to having very low frequency (low voltages) and very long wavelengths (often miles across). You also understand that microwaves and heat can generate motion. The divisions of force from all field activity is not hard and fast, but flexible, and appears as a gradient. You are familiar with how a small fast gear can drive a large slow gear in order to actuate power. Tori around a heavy flywheel will sustain forceful action. It is the job of an engineer to understand how to reestablish one field equation with another in order to formulate the needed force. Yet mechanical force and alpha waves are both born of equations of the ST Fields of the electromagnetic spectrum.

When you hear the word "momentum," you think of sustained action and time. The expression of T Field is time as you know it. As T Field slows into a potential point its expression of time graduates into an expression of force. A still point in time is a complete condition of force as resistance. The slowest rate that time can express is the most resistance (most pulling force) it can express. At a still point the T Field is sustained power, like an untapped reserve of energy. And yet the idea of a potential point as sustaining indicates an expression in itself which is unavoidable in the context of a universe. Without a spatial sphere, an S Field balance, the point of potential force has no definition as such. It must engage something else or not exist at all. Beyond the 'point with an address in space' we cannot speak.

Time tends toward linearity and axialarity. Because of this characteristic, the ultimate undertone (accompanied by the ultimate overtone) becomes available as corpuscular radii throughout the pressure systems of the universe. The literature of the Hindu Upanishads refers to the universal design as pearls on a necklace. This visual metaphor is a true depiction of points of T Field in a cyclic expression containing (as a torus) the pressures of the S Fields. The equation between the push (action) and pull (restraint) from a point fulcrum is the measure of your force in tension, in balance.

A point fulcrum for an equation is the USUT Fields extending balance to the ST Fields. That fulcrum centers a large overtone which describes an environment, a universal pressurized space in which the ST Field finds its balance. When time is expressed slowly, for instance as the atoms in a natural piece of iron ore, you have a long cycle of existence. The integrity of the rock, one atom in slow motion with regard to its neighboring atom, constitutes one speed, while the rock itself, traveling with the earth at universal speeds, has a compounded speed. The measure of force is always a matter of a handful of conditions: what with what, how and when, and how many. Your definition of "mass" can be equally as variable. All forces, both potential and kinetic, are described by sets of tori in combination, and how those tori might undertone or overtone.

Force must be defined as S Field coming up against T Field and T Field coming up against S Field and confining it. Force is not in evidence without opposition. When the S Field or the T Field is expressed in harmonic tori, you can see forces at work in direct relationship to motion, especially slow motion.

This word "magnetism" is badly assigned. You have noticed that we use the word only to help you identify a familiar condition. Magnetism has meant to you 'attraction and repulsion'. We want you to look at the phenomena in different ways.

When matching magnets are placed together they will overtone all their tori and mechanically pull together. The pulling inward into spherical tensions of overtones is an act of CLOSURE. Tori around moving lines of field act like a rubber band. When overtoning takes place between tori you can notice an increased potential center point for that overtone. All radii of that overtone increase their tension, pulling inward to center, exhibiting CLOSURE. An increased surface tension contains pressures of S Fields swirling within. The surface tension of tori around a magnet does not describe a sphere but an apple shaped torus, yet it has one center point. When two small magnets are aligned and overtoned, the inward pulling force is sufficient to draw the magnets together with mechanical motion. It is as though a rubber band was stretched between them. Then the rubber band CLOSES to center.

The tensions of unity in the T Field called CLOSURE, having been enacted from two sources, 1) the integrity of the iron bar, and 2) the separated fields drawn through the bar, show an increase by way of tori which draw upon the energy resources of the fields at large. The atomic structure of iron has the unique ability to separate fields into its S and T Field components (like dividing one road in half with a yellow line to direct two way traffic). The spinning of tori around a *moving magnet* add to the mechanics of CLOSURE by adding a slow speed tori around the field flow.

A magnet sitting still on a table, has a tori spin always, simply because it is rotating with the earth and traveling in space. That is its basic tone (or overtone). All other tori spins will be relative to that basic tone. Any additional motion or flow will be an increase to that. And that basic tone can be compared to pressure within a universal sphere; it is multidirectional and omnipresent. There is a direct relationship between ratios of speed and ratios of pressures. This deserves careful study.

When two small magnets CLOSE together in overtone, it takes a little time. It takes a little time for tori to recognize one another, to engage the fields and to adjust the potentials and speeds of a new overtone. Motion, even a little bit of motion, reduces the differences in speed between the flux field flows and the tori formed by motion, thereby reducing the resistance to CLOSURE. As resistance is reduced, motion increases, reducing resistance further. In this way, CLOSURE accelerates, once begun.

The same process of tori speed escalation applies to falling objects in a gravitational field. You may have noticed that when an astronaut in a space station drops his wrench, it does not accelerate away from him, but floats about in space. Falling objects in a gravitational field accelerate in that field toward a central point of CLOSURE. The GT Field lines are flowing fast like a river. A dropped object is in resistance to that GT Field flow because its tori is spinning slowly. As it moves in the direction of the GT Field flow, the object's tori is in a position to overtone with tori around the GT Field lines. Overtoning takes time. With each increment of its fall, the tori of the object speeds up, offering less resistance to tori speeds of the GT Fields. The object falls at an accelerating rate until it reaches the speed of the flow of the GT Field. Before you can measure a cessation of acceleration, the object has hit the ground.

If an object moves perpendicular to the direction of the GT Field, its slow-moving tori (like a child's waterwings) keep it afloat. The resistant tori spin in a position that prevent overtoning from taking place. The faster the object's tori spin moves in opposition to the gravitational pull, the less opportunity there is for overtoning, and in fact, there is opportunity for pushing away and out. This is simple engineering for flying craft once you know how to manipulate S Field predominant tori. For instance, tori around your helicopter rotor blades not only push air but limit the gravitational effects upon the blades, but not upon the body of the craft. Learning to create powerful S Field tori will take you into space.

ILLUSTRATION 9-14
The stone detail of the feathered serpent at the temple of Quetzalcoatl at Teotihuacan, Mexico shows us
that the images of the Gods of Central America represented cosmic principles.
The serpent with the feathers curled back on its head illustrated the electric aspect of the ST Fields.

This two headed serpent from Monte Alban, Mexico illustrates
the equal and opposite waves of forces that carry messages from
the priests to their Gods.

CHAPTER 7

OPENING THE PETALS OF COMPASSION

A DYNAMIC LESSON

A condenser, or capacitor, is a little device that stores up energy. It consists of two T Field plates (usually metal conductors) placed facing each other with a space (or a dielectric material) in between. When one plate is charged, the opposite plate shows an equal and opposite charge. What happens in between?

When the source of the electric charge is stopped, the charges on the plates retain their original charge for an extended time. It is as though energy had accumulated in the middle space then slowly feeds energy back into the plates, demonstrating that an energy gain has taken place. This familiar device is commonly used in electronic equipment to regulate sensitive power flows. It is used for energy storage in many mechanical – electric devises.

What happens in between the plates, in the air or dielectric material, is that a strong S Field predominance develops, equivalent to the T Field predominant plates, and with charges compatible with its adjacency to the plates. The S Field space accumulates pressure and energy gain which is slowly released.

In addition to what is seen and measured are two invisible spheres, one in overtone around each plate, with appropriate charge predominance. Those two field spheres join in a planar cord at the center of the dielectric area. On this plane an S field sphere locates at the center of the cord, sharing common axes with the outer spheres. This center S Field sphere manifests pressure and energy gain. **Illustration 7-1** will show that on axis AB a spiral begins outward (and inward). Upon this spiral any signal events are recorded in a high energy state. And, on this axis the USUT Fields becomes strongly manifest, doubling the energy of the lesser fields. When the current build-up stops, the energy of the dielectric feeds back into the quiescent circuit at a given rate. The rate of feedback tells you much about *time* as relates to your lower fields.

ILLUSTRATION 7-1

THE CAPACITOR

The space between two electrically charged metal plates is filled with S Field energy which is equal and opposite to the charge on the plates. It takes the form of a sphere in overtone. There are also overtoned spheres around each charged metal plate. The three spheres together share a common central plane. On this plane energy spirals in and out as a result of any oscillation in the fields nearby. Any motion on this plane causes tori which broadcast outward into the fields at large.

The capacitor is used in electronics because it holds and feeds back energy that was given to it by the charged plates. It therefore act to regulate the flow of energy in a circuit. At the same time it acts to broadcast signals across fields. Used with an oscillator or a pulser it can become a carrier wave that includes finer messages. It is like a miniature broadcasting station.

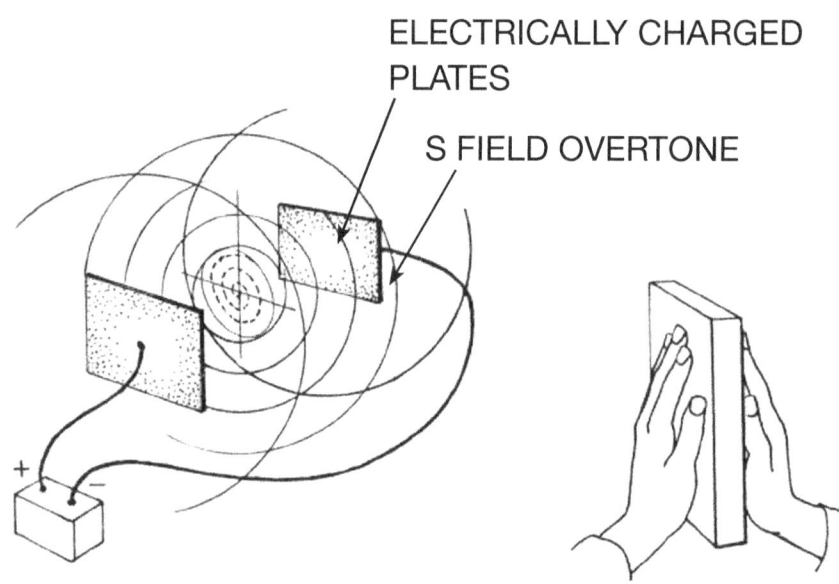

Placing a cardboard box or envelope between the palms of the hands the box and its contents can be broadcast outward. If the person in prayer can hold the image of his prayer clearly in his mind or 30 seconds and then also the recipient of the prayer clearly in his mind for 30 seconds he may effectively transmit the message. His fingertips should line up. He may use a color to help the vibration of the carrier wave. This is recommended for loving prayers of healing. To address the Spirit of God he may face east. To contact a human being he may face west.

You will notice that the position of condenser plates is very much like a situation for broadcasting. Charges can jump back and forth between the plates. But because of the presence of the dielectric, charges are held hostage upon their own plates. The total potency of the charges can be varied periodically and the feedback from center stabilizes the frequency patterns.

However, the radio wave signals are regulated and distributed evenly by the dielectric space which sends broadcast signals out into the atmosphere. Charges, traveling back and forth between the two terminals of the plates stimulate radio broadcast in the fields at large. The presence of the strong USUT Fields doubles the amplitude of signals as fields send forth broadcast waves.

Now we can compare this phenomena to human practices. When you were a child in Sunday school you were taught to place your hands together palm to palm, fingers to fingers, in a gesture for prayer. You can see the same positions of the hands in ancient drawings from all cultures. It is a traditional way to pray to higher energies. Two hands, placed together is, in fact, a condenser. In that practice the USUT Fields are directly impressed with the message of the prayer.

If the hands in prayer do not touch, encouraging the charges to leap a space, you will have an improved condenser and broadcasting apparatus. With the insertion of a dielectric between the hands, such as a paper or an empty cardboard box, you can feel more relaxed. In addition, a message or picture on or in a box can translate clearly within the context of your prayer.

The position of the hands in proximity to the head and various chakras can carry a message from that bodily place. For instance, placing medicinal plants and flowers in your prayer box, and holding your hands in prayer near your heart, and visualizing compassion around a certain brother or sister, you can convey and direct a healing to them at great distances. You become a living radio which sends organized information that their minds can perceive (although it may be unconsciously perceived). the intensity of your compassionate desire in prayer will be directly reflected in vthe extent and potency of your broadcast signal.

It may have occurred to you that the USUT Fields, the grand overtones of the ST Fields, are intimately concerned with human and earthly activity. Any action is doubled, any signal is overtoned and magnified into principles and laws. It may have occurred to you that the USUT Fields are the dwelling place of living beings, and from the beginning you have called them angels, God, goddesses and relative figures. It is in truth the reality realm of higher beings that know who you are and what you think and do, and who are usually willing to speak or act in your behalf. That is what prayer is all about. The human condenser of hands together helps to communicate to those realms. Although some people may reject that idea as preposterous, we are here beside you to tell you that it is true.

BROADCASTING IN RINGS OF COMPASSION

In our lesson now we ask that you visualize rings of energy radiating out from you to great distances. Your rings, intersecting with rings of other beings, human, plant, mineral, device, and all the little dwellers of earth, qualify life forces with the compassionate nature of your own being. Your rings can assist the love of life in all they intersect.

Let us enlarge upon that practice. there are two halves of your body with their own predominances. Each half is equal and opposite in predominance, such as in electrical-nerve charge, in male-female, in S and T Field, in right and left brain, etc. The spinal column centers the two halves, and as well, some organs, some specialized glands, the heart, and parts of the brain like the pineal gland and the thyroid.

The axial plane in the center of your two-sided body is a place where two spheres intersect. On this plane the physical and spiritual energies of your body merge. Interaction and energy exchanges between the two sides of your body, by way of the centering Universal Field, broadcast outward in great spheres. Your thoughts and the qualifying signals of your being are transited outward to infinite distances. These are your personal signals that we can hear and interpret.

It will help you and all others if you can hold the tones, the harmonies, of good will and compassion in your emissions. You have a primary ring around you, which, like your home, may get dusty from time to time. The first ring is also where your surface tension builds. It is a good visualization to clear and clean that first ring and place flowers within it, to wash it down with color, and make it sparkling clear with reflections of rainbow light.

Clear your rings of sad memories and desperate fixations with the waters of compassion. Wash with gentle tears the tragedies of loss.

Then, with fires of white and violet flames, intensify the Universal Fields of compassion within your own being. As you breath deeply inward, raise your hands above your head in prayer position with fingers parted. As you exhale deeply bring your hands and arms out to your sides to receive your blessing. Fill your rings with the swirling beauty of compassion to radiate outward, knowing it will be a blessing to all it encounters, reaching infinitely into the starry skies, and plunging weightlessly into the depths of the planet. As you wholeheartedly engage your energy, know that the Universal Energy will match your outreach, one on one, in amplification.

Within the realms of the Spirit there are beings devoted to energies that support he laws of Equivalencies, and of Renewal, and of Devotions toward the whole. These Beings, when asked, can amplify and support in structure your own realization and projection of the greater LAWS.

> **The Universal lives can alter your biology to allow for your outreaching into the compassionate realms of the cosmic community.**

The unfolding of the petals of the flower of compassion is the opening of the door to enlightenment, the Christed consciousness. It is the complete change between the animal humanity and the spiritual humanity. In this state you will feel weightless and well, and your radiance will restore all the lives around you. Bringing the fullness of the Universal Fields into the chalice of your body, all its organs, skeleton, brain and every single cell, prepares the way for the cosmic gatekeepers to unlatch the barrier doors between a lesser reality and a far greater reality. Do not be a flake or a slouch! It requires an energetic willingness to interchange and focus, in alert meditative exercises, everyday, in quiet or in full activity. Then assistance will come.

Remember that a sudden stop by your body will throw the USUT Fields out front. That is the nature of shock. A person who strikes or is stricken, who hits or is hit, momentarily looses their centering spiritual life. They become unconscious and may even act unconsciously. A stumbling block may cause a fall and a spiritual life is unsettled. Spiritual mental control of the human body by way of biological contact points is what separates animal life from human life. Movements of harmony and smooth rhythm can spiral spiritual essence inward to amplify the spiritual human centers. Religious dances came about naturally. The space tori around a moving body swirls in confirmation of harmonic order. Interconnecting action in the fields surrounding dancers are carried like radio waves into the space continuum. Specific gestures carry encoded messages to convey recognition and honor to a higher Being. Spinning prayer wheels used by Tibetans broadcast with circling alternating currents. A strong S Field tori can develop a strong US Field center sphere, and a US Field will significantly amplify any field activity.

Amplification of signaled messages always comes about by a series of signal exposures to S Fields and US Fields. Compassion is an attitude which is the epitome of US Field expansion and energy gain. Compassion carries Renewal with it in accelerated growth and repair. Compassion is a higher energy which overlays a situation, a place, a life, a land and its people. Those beings who amplify compassion stand in series in any realm, standing at their chosen post to carry the light of love to greater heights.

There is a story, and folks swear it is true, about an abandoned lighthouse that stood on the banks of Lake Huron since 1830. After years of service it was replaced by a nearby larger, more modern lighthouse. Its old bulb and wires were removed, leaving the big French lenses in place. The lighthouse was restored for tourists without its light. New keepers kept it up. But, after the new keeper died a ghostly light began to burn brightly in the tower on many foggy nights. The neighbors all saw the light and investigators found no good explanations. The keeper's widow, living there, knows it is her husband's ghost who makes a light glow in the lighthouse he loved and maintained in life, a light that reaches out to lost sailors beyond the rocky shore. There is a moral to this story. Devotion to the compassionate light makes all good things possible.

A series of enlarging lenses in a microscope makes a light image bigger, step by step. Something that was too small to see comes into focus in the eyepiece amazingly clear and large. By the subsequent passing of tender caring from person to person, minute to minute, person to animal, animal to plant, plant to earth, earth to sky with its nourishing rain, a loving beginning spreads out over the land. The seeds from one stalk of wheat can make a whole field of wheat. A tiny brown acorn develops into the many branches of a tall oak tree. Stone by stone, cathedrals are built. A body can be thrilled hearing the blended voices of a choir singing praises unto heaven. Each act of love and compassion builds upon the first. The great living Beings that stand strong as cornerstones of compassion will always serve with explosions of love to fill the hearts of all who await in joy, in freedom from isolation, and limitation. Compassion will always show energy gain.

Freedom is enhanced by democracy. Democracy is upheld by education. There is no freedom without education for all. Education is the backbone of any advanced society where love and compassion can thrive under law and where law can be rested upon the fulcrum of compassion. The circle of civilization balances on the human ability to amplify cosmic love.

When a baby is born from its mother, it is just beginning to grow. the amplification of compassion as educational practice causes the body and brain of the child to grow in a very special way. The child's biology adapts to its energy presence. A baby raised by wolves will remain a wolf. A baby raised in a complex civilized society will have the capacity to flourish in that place and time. Cells change and neurons extend their avenues to accommodate. Genes undergo a change. Social behavior is modified, not only by teaching, but by biological adaptation.

People do not automatically grow up with the capacity to carry forth a spiritually empowered civilization. Their delicate senses are developed by caring and teaching. A child who has not been properly educated by the age of twenty is much like an aborted fetus, incapable of further growth. There are many countries on your planet where most of the children are in that condition. Everyone must take whatever opportunity and expression they may have to learn and to teach that which is spiritually inspired. With increases in population the problem of child development grows exponentially. Crisis and panic are the results of educational disparity and lack. Not one person can sidestep an obligation to the spiritual growth of children over the whole world.

The simple lessons of faith, the practices offered here, should not be taken to be like a game. They describe a means to growth, adaptable to any known religion of GOD. Spiritual growth is your only road to take to know the meaning of freedom, the only road to a successful advanced technical civilization. Obligations to the whole always begin with the education and training of yourself and your family.

As you are writing these paragraphs, the educators of the country of Tibet have been either killed by Chinese invaders or have escaped into the crowds of other races. In that country the only educators were monks. The invaders are not supplying teachers, schools or hospitals in sufficient numbers to the Tibetan people who remain in their homeland. The educated monks, looking back at the conditions in their homeland, suffer deep remorse to know that their families are so oppressed. They know that when educators are gone the children cannot take any place in a modern world. They become slaves to a master. When the light of compassion and education goes out, a civilization will fall into collapse. When the nourishment of the spirit ceases, the children do not grow. It is necessary that parents, teachers, and political leaders realize that the ultimate development of civilization depends upon how well their teaching stimulates self empowerment in children. Success requires individual freedom and self-willed responsibility. Human success requires a freewill personal decision to access the lawful powers of ALL THAT IS.

What you know and what you do personally is critical to the whole planet. How you work and think in the spiritual silence of your own home is signaled around the world. Your personal life counts even to the outer edges of our galaxy. One person with a will to compassion makes a qualified difference in every molecule everywhere. Although it is hard to grasp, it is true that there is no private world. There is ONE overall truth which every being everywhere must share to a fullness. The Star People stand side by side with you in the light of one compassionate truth which we all strive to comprehend.

> **ONE COMPASSIONATE TRUTH is the vortex from which all love and sharing outpour. It is the well-spring of manifestation and the very nature of your creative gifts of life.**

Wars among races and peoples have always come about by the provocation of want. In a limited world there is never enough to go around with an equal share for all. You are beginning to understand that there are ways and means to help people get what they need and want. Sacrifice is not the answer to adjusting to a limited environment. Learning and changing is the better road.

The peoples of mesoamerica in olden times practiced sacrificial slaughter on the alters of their temples. From your view today, it seems savage to kill fine men and women to buy favors from gods. In times before A.D. 1000, during the civilizations of the Zapotec, Toltec, Aztec, and Maya, the people recognized that gods dwelled within their body. Many individuals looked upon death as a privilege of human release and volunteered themselves for sacrifice. It was their way to show cosmic love. It was necessary that the martyrdom of humans and animals come to an end. The deaths were distorting the experience of life. Today, we advocate freedom as a means to make the experience of life meaningful. In the long continuity of time the Spanish invasion of mesoamerica was spiritually important. The compassion of the Christ was eventually influential in bringing value to human life as well as the life of the gods of olden times.

Today most everyone in the world can own a small transistor radio and hear music of their choice. What a miracle that is! People can see and hear new friends in every country via computer and TV. Soon, people will develop healthy diets from foods that do not ruin lands and deprive others, etc. People can adapt in a comfortable way to living on a limited planet. The psychological panic of deprivation is not necessary, and not appropriate to a compassionate way of life.

FAR OUT AND CONNECTED

The human body is a marvelous invention. It was created as a finely tuned, articulate, electronic biotronic instrument. Your senses coordinate a series of wavelengths, thus overtoning to new dimensions. You manipulate a compendium of active chemicals that facilitate motion, and recognition, and creative enterprise. Your soul learns as you experiment with your daily experience. You grow from your interaction with other living matter, from interpreting the signaled impulses within the cosmic fields. Now, as before, you are asked to keep a continuous thread of consciousness throughout your embodiments. Always keep contact with WHO you are.

The fields carry the laws as musically toned signals. You have direct access to those laws. Some have called this access a "conscience" or "with-science." If you look deeply within yourself, pushing aside what is superficial, you will recognize those laws. Human progress is evident when cosmic laws stand behind every human thought and activity. As you are a Divine Manifestation you are asked to fully recognize the presence of The Divine within yourself, and within all others. You are asked to be Divinely Harmless. Cosmic divinity will never advocate or support the interruption of manifested life.

It will serve you to learn the functions of each of your sense organs, both physical and etheric, in order that you may expand their ranges by practice. Expand your talents . Reach into the cosmic reservoir of the arts. In every daily activity believe in your greatest connectedness.

Your body is certainly connected, cell to cell, bone to bone, synapse to tendon. Your spirit self is connected to the body you have chosen. Yet your connectedness can easily change as you, as a complete entity, find the need to change. Your DNA does not have a permanent hold on you. You are in charge, and you have the ability to alter your DNA. It takes time and cooperative practice. As you know, your scientists today are altering DNA of living things in their laboratories. Good medicines may develop from their studies. But, do not depend upon others to work miracles on your behalf. Life changes come from you.

There is a purpose to your planetary life. It is to allow yourself the experience of working with specific energy levels, and to express your freewill in wise decisions. Young children make important small decisions. Adults make decisions that have broader importance. As a cosmic citizen your will make far reaching decisions based upon your extended education.

You are cultivating your own being. Over millennium you will evolve according to your own choices of WHO you want to be. No master will drive you like a slave, this way or that way. You will make decisions on the basis of your own experience and education. Like a two-way radio, you will be able to interfunction on the cosmic wavelengths with all antennae up and buzzing.

HEARING THE RIPPLES ON THE WATERS

To be a receiver of channeled or gifted energy you will be like a finely tuned radio that finds the station of choice and amplifies the signal into sound with hi-fidelity perfection. An old fashion radio had an engaging tuning device that you changed with your dialing to the correct shape to match a selected wavelength. You chose your favorite station, then the signal had to proceed through methods of amplification, and then to a magnetic speaker membrane so that your ears could hear the sound which the signals had intended. Remember "Hi Ho Silver!," and the Boston Symphony Orchestra? As a child in the old days it all seemed perfectly natural while your parents were amazed and shocked. At the turn of the last century, if you had told a man that his room was full of talking radio voices he would have thought you were mad. What he could not hear or see certainly could not exist. Without the little wooden box that plugged a wire into the wall, a person was left to the silence of the room. The radio box was partly responsible for peoples adoration of magical gadgets.

Then, as now, you live in darkness in the midst of light. You are blind in a world full of swirling colored energies of the higher octaves.

Maybe if you found some magical sunglasses to put on your nose, or mystical pills, or went into weird ecstasies in a smoky room, it would all be there for you. As your heart goes out to a blind man on a street corner, our hearts go out to you. This time your gadgets will not work. To access the beautiful higher octaves, you must train your whole being to see beyond the heavy curtains of your human-animal wavelengths. You must practice in order to alter the biology of your brain and the etheric organs of your body. This can only happen when you truly desire to know what is above and beyond. Practice amplifies the loving presence. You will feel it, know it. Long wavelengths take a lot of time.

Here you are now, relaxed and comfortable in your special place. The petals of the flower of the Compassionate Light flutter in the gentle winds of summer. You have swept away the tensions and spasms from your entire cellular being. You give the higher dimensions permission to enter your body. You become a willing receptacle in trust and faith of the ultimate perfection of the ALL. You become as a little child, letting go of all your burdens and smiling in the love of life. Your heart is light as a white feather and shining with the love all around you. The permeating energy of the Compassionate Being knows WHO you are, blessing you with perfect LOVE.

Like the strings on a guitar, there are nerves in your body of special length and delicacy that are designed to vibrate at particular frequencies when under appropriate tensions. Unlike a guitar, these nerves are living substance that can respond to the Universal Field energies. The long wavelengths of the US Fields, alive with signaled information, can find those nerves in a condition of resonation. Your nerves need to be like tuning forks in perfect shape and tension to accommodate the transiting signals of the compassionate spirit. When your nerves intercept those signals, they develop "standing waves" that amplify and step-down the signals, to undertone the information. Amplification of Compassionate Being is the full complete message. Your being will resonate outward an amazing grace without words. These "standing waves" are not feelings of personal power, they are feelings you might experience swimming in a perfumed pool with many friends and white lilies floating about your head. You will feel buoyant and well. Then you will want to tell the whole world about the meaning of beauty. Questions that lay painfully buried in your being for decades will surface and be resolved. Disturbing questions on your mind will be addressed and brought to rest. Once you have reached a practiced level of communication, special beings will answer technical questions that are formulated with care and accuracy. All these things will come about over a period of time and practice.

The Overlighting Beings whom you are receiving know that in your human condition you can only adjust slowly to change, that you grow into understanding as your biology changes slowly. Therefore, they will help you only as you ask for help, and only as your being is capable of receiving. Once you have experienced the great Beings who assist the Compassionate Spirit, you will want to spend more and more time in their presence just as lovers desire to spend more time together in the garden of their love. Soon they will never leave you, even in your dark and stressful moments. Your overlighting will be constant and complete. Your communications will span the octaves and you will be glad. All of these things will come about only through your personal desire and individual freewill as you communicate one on one with your galactic companions.

The Tibetan White Goddess of Compassion
illustrates for mankind a principle of physics,
that is, the merciful balancing justice
of the Universsal Fields.

www.ingramcontent.com/pod-product-compliance
Lightning Source LLC
Chambersburg PA
CBHW081229080526
44587CB00022B/3869